PHASE TRANSITIONS
Modern Applications
2nd Edition

PHASE TRANSITIONS
Modern Applications
2nd Edition

Moshe Gitterman
Bar Ilan University, Israel

World Scientific

NEW JERSEY • LONDON • SINGAPORE • BEIJING • SHANGHAI • HONG KONG • TAIPEI • CHENNAI

Published by

World Scientific Publishing Co. Pte. Ltd.

5 Toh Tuck Link, Singapore 596224

USA office: 27 Warren Street, Suite 401-402, Hackensack, NJ 07601

UK office: 57 Shelton Street, Covent Garden, London WC2H 9HE

Library of Congress Cataloging-in-Publication Data
Gitterman, M.
  Phase transitions : modern applications / Moshe Gitterman. -- 2nd edition.
  pages cm
  First edition (2004) has title: Phase transitions: a brief account with modern applications.
  ISBN 978-9814520607 (hardcover : alk. paper)
  1. Phase transformations (Statistical physics) 2. High temperature superconductivity.
3. Superconductors. I. Title.
  QC175.16.P5G58 2013
  530.4'74--dc23

                                                                            2013033570

British Library Cataloguing-in-Publication Data
A catalogue record for this book is available from the British Library.

Printed in Singapore

Preface to the Second Edition

In spite of its century-long history, phase transitions remain a central problem in many-body theory. Because of the crucial importance of long correlations, the details of the structure and properties of different substances count for very little in the many-body properties of different systems in Nature. The theory of phase transitions explains why many different systems, classical and quantum, solids, liquids and gases, all exhibit similar properties near the phase transition. An examination of the singularities of physical parameters at the critical point allows one to understand the characteristic properties of these systems, much as a person's character manifests itself most conspicuously under critical circumstances. The theory of phase transitions and critical phenomena is used extensively in modern science and technology.

These facts explain the appearance of the new books and new editions of already published books on phase transitions. It is the reason for this new edition of the original 2004 book. The most important new material in the second edition is a detailed analysis of the phase transitions in chemically reacting and moving systems, as well as an analysis of the thermodynamics of phase transitions. The appearance of the second edition gave me the opportunity to add some sections and to introduce changes to improve the clarity of the text.

The book can be used as a textbook in a course on phase transitions, as well as an introduction for graduate students in other disciplines who work with phase transitions.

Preface to the First Edition

This book is based on a short graduate course given at New York University and at Bar-Ilan University, Israel. The decision to publish these lectures as a book was made, after some doubts, for the following reasons. The theory of phase transitions, with excellent agreement between theory and experiment, was developed some forty years ago culminating in Wilson's Nobel prize and Wolf prize awarded to Kadanoff, Fisher and Wilson. In spite of this, new books on phase transitions appear each year, and each of them starts with the justification of the need for an additional book. Following this tradition we would like to underline two main features that distinguish this book from its predecessores.

Firstly, in addition to the five pillars of the modern theory of phase transitions (Ising model, mean field, scaling, renormalization group and universality) described in Chapters 2–5 and in Chapter 7, we have tried to describe somewhat more extensively those problems which are of major interest in modern statistical mechanics. Thus, in Chapter 6 we consider the superfluidity of helium and its connection with the Bose–Einstein condensation of alkali atoms, and also the general theory of superconductivity and its relation to high temperature superconductors, while in Chapter 7 we treat the x–y model associated with the theory of vortices in superconductors. The short description of percolation and of spin glasses in Chapter 8 is complemented by the presentation of the small world phenomena, which also involve short and long range order. Finally, we consider in Chapter 9 the applications of critical phenomena to self-organized criticality in scale-free non-equilibrium systems. While each of these topics has been treated individually and in much greater detail in different books, we feel that there is a lot to be gained by presenting them all together in a more elementary treatment which emphasizes the connection between them. In line with this attempt to combine the traditional, well-established issues with the recently published and not yet widely known and more tentative topics, our fairly short list of references consists of two clearly distinguishable parts, one related to the classical theory of the sixties and seventies and the other to the developments in the past few years. In the index, we only list the pages where a topic is discussed in some detail, and if the discussion extends over more than one page then only the first page is listed.

We hope that simplicity and brevity are the second characteristic property of this book. We tried to avoid those problems which require a deep knowledge of specialized topics in physics and mathematics, and where this was unavoidable we brought the necessary details into the text. It is desirable these days that every scientist or engineer should be able to follow the new wide-ranging applications of statistical mechanics in science, economics and sociology. Accordingly, we hope that this short exposition of the modern theory of phase transitions could usefully be a part of a course on statistical physics for chemists, biologists or engineers who have a basic knowledge of mathematics, statistical mechanics and quantum mechanics. Our book provides a basis for understanding current publications on these topics in scientific periodicals. In addition, although students of physics who intend to do their own research will need more basic material than is presented here, this book should provide them with a useful introduction to the subject and an overview of it.

Contents

Chapter 1

Phases and Phase Transitions

In discussing phase transitions, the first thing that we have to do is to define a phase. This is a concept from thermodynamics and statistical mechanics, where a phase is defined as a homogeneous system. As a simple example, let us consider instant coffee. This consists of coffee powder dissolved in water, and after stirring it we have a homogeneous mixture, i.e., a single phase. If we add to a cup of coffee a spoonful of sugar and stir it well, we still have a single phase — sweet coffee. However, if we add ten spoonfuls of sugar, then the contents of the cup will no longer be homogeneous, but rather a mixture of two homogeneous systems or phases, sweet liquid coffee on top and coffee-flavored wet sugar at the bottom.

In the above example, we obtained two different phases by changing the composition of the system. However, the more usual type of phase transition, and the one that we will consider mostly in this book, is when a single system changes its phase as a result of a change in the external conditions, such as temperature, pressure, or an external magnetic or electric field. The most familiar example from everyday life is water. At room temperature and normal atmospheric pressure this is a liquid, but if its temperature is reduced to below 0°C it will change into ice, a solid, while if its temperature is raised to above 100°C it will change into steam, a gas. As one varies both the temperature and pressure, one finds a line of points in the pressure–temperature diagram, Fig. 1.1a, along which two phases can exist in equilibrium, and this is called the coexistence curve.

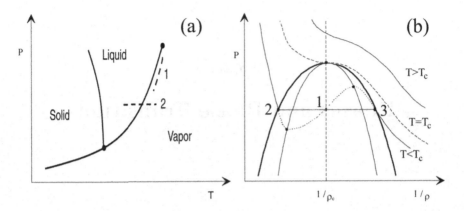

Fig. 1.1   (a) P–T and (b) P–v phase diagrams, where $v = 1/\rho$ is the specific volume.

We now consider in more detail the change of phase when water boils, in order to show how to characterize the different phases, instead of just using the terms solid, liquid or gas. Let us examine the density $\rho(T)$ of the system as a function of the temperature T. The type of phase transition that occurs depends on the experimental conditions. If the temperature is raised at a constant pressure of 1 atmosphere (thermodynamic path 2 in Fig. 1.1a), then initially the density is close to $1\,\mathrm{g/cm^3}$, and when the system reaches the phase transition line (at the temperature of 100°C) a second (vapor) phase appears with a much lower density, of order 0.001 $\mathrm{g/cm^3}$, and the two phases coexist. After crossing this line, the system fully transforms into the vapor phase. This type of phase transition, with a discontinuity in the density, is called a first order phase transition, because the density is the first derivative of the thermodynamic potential. However, if both the temperature and pressure are changed so that the system remains on the coexistence curve (thermodynamic path 1 in Fig. 1.1a), one has a two-phase system all along the path until the critical point (for water $T_c = 374$°C, $p_c = 220$ atm.) is reached, when the system transforms into a single ("fluid") phase. The critical point is the end-point of the coexistence curve, and one expects some anomalous behavior at such a point. This type of phase transition is called a second order one, because at the critical point the density is continuous and only a second derivative of the thermodynamic potential, the thermal expansion coefficient, behaves anomalously. Anomalies in thermodynamical quantities are the hallmarks of a phase transition.

In Fig. 1.1b, the coexistence of the two phases is shown in the pressure p–specific volume (inverse density ρ) phase diagram. A state of the system is shown by point 1 in Fig. 1.1b, which describes the mixture of liquid and vapor located at points 2 and 3, respectively. The small decrease in density (at constant temperature and pressure) means that the amount of vapor in the system has increased while that of liquid has decreased. The energy absorbed by system during the compression is expended as latent heat at the liquid–vapor transition. For pressures larger than the critical pressure, the system is homogeneous.

Phase transitions, of which the above is just an everyday example, occur in a wide variety of conditions and systems, including some in fields such as economics and sociology in which they have only recently been recognized as such. The paradigm for such transitions, because of its conceptual simplicity, is the paramagnetic–ferromagnetic transition in magnetic systems. These systems consist of magnetic moments which at high temperatures point in random directions, so that the system has no net magnetic moment. As the system is cooled, a critical temperature is reached at which the moments start to align themselves parallel to each other, so that the system acquires a net magnetic moment (at least in the presence of a weak magnetic field which defines a preferred direction). This can be called an order–disorder phase transition, since below this critical temperature the moments are ordered while above it they are disordered, i.e., the phase transition is accompanied by symmetry breaking. Another example of such a phase transition is provided by binary systems consisting of equal numbers of two types of particle, A and B. For instance, in a binary metal alloy with attractive forces between atoms of different type, the atoms are situated at the sites of a crystal lattice, and at high temperatures the A and B atoms will be randomly distributed among these sites. As the temperature is lowered, a temperature is reached below which the equilibrium state is one where either the atoms A and B are separated or the positions of these atoms alternate, so that most of the nearest neighbors of an A atom are B atoms and vice versa.

The above transitions occur in real space, i.e., in that of the spatial coordinates. Another type of phase transition, of special importance in quantum systems, occurs in momentum space, which is often referred to as k-space. Here, the ordering of the particles is not with respect to their position but with respect to their momentum. One example of such a system is superfluidity in liquid helium, which remains a liquid down to $0\,\mathrm{K}$ (in contrast to all other liquids, which solidify at sufficiently low temperatures and high pressures) but at around $2.2\,\mathrm{K}$ suddenly loses its viscosity and so acquires very unusual flow properties. This is a result of the fact that the particles tend to be in a state with zero momentum, $k = 0$, which is an ordering in k-space. Another well-known example is superconductivity. Here, at sufficiently low temperatures electrons near the Fermi surface with opposite momentum link up to form pairs which behave as bosons with zero momentum. Their motion is without any friction, and since the electrons are charged this motion results in an electric current without any external voltage.

Phase transitions occur in nature in a great variety of systems and under a very wide range of conditions. For instance, the paramagnetic–ferromagnetic transition occurs in iron at around $1000\,\mathrm{K}$, the superfluidity transition in liquid helium at $2.2\,\mathrm{K}$, and Bose–Einstein condensation of atoms at $10^{-7}\,\mathrm{K}$. In addition to this wide temperature range, phase transitions occur in a wide variety of substances, including solids, classical liquids and quantum fluids. Therefore, phase transitions must be a very general phenomenon, associated with the basic properties of many-body systems. This is one reason why the theory of phase transitions is so interesting and important. Another reason is that thermodynamic functions become singular at phase transition points, and these mathematical singularities lead to many unusual properties of the system which are called "critical phenomena". These provide us with information about the real nature of the system which is not otherwise apparent, just as the behavior of a poor man who suddenly wins a million-dollar lottery can show much more about his real character than one might deduce from his everyday behavior. A third reason for studying phase transitions is scientific curiosity. For instance, how do the short-range interactions between a magnetic moment and its immediate neighbors lead to a long-range ordering of the moments in a ferromagnet, without any sudden external impetus? A similar question was raised (but not answered) by King Solomon some 3000 years ago, when it was written (Proverbs 30: 27): "The locusts have no king, yet they advance together in ranks".

Chapter 2

Thermodynamics of Phase Transitions

2.1 Classification of Phase Transitions

The description and analysis of phase transitions requires the use of thermodynamics and statistical physics, and so we will now summarize the thermodynamics of a many-body system [1]. In thermodynamics each state of a system is defined by some characteristic energy. If the state of the system is defined by its temperature T and its pressure P or volume V this energy is called the free energy. One part of this energy is the energy E of the system at zero temperature, while the other part depends on the temperature and the entropy S of the system. If the independent variables are the temperature and pressure, then the relevant thermodynamic potential is the Gibbs free energy $G = E - TS + PV$, while if they are the temperature and volume it is the Helmholtz free energy $F = E - TS$. The differentials of these free energies for a simple system are

$$dG = -S dT + V dP, \qquad dF = -S dT - P dV. \qquad (2.1)$$

If the system has a magnetic moment there is an extra term $-M dH$ in the above expressions, and if the number N of particles is variable we must add the term μdN, where μ is the chemical potential. Then the first derivatives of the free energy give us the values of physical properties of the system such as the specific volume ($V/N = [1/N]\partial G/\partial P$), entropy ($S = -\partial G/\partial T$) and magnetic moment ($M = -\partial G/\partial H$), while its second partial derivatives give

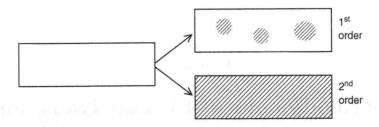

Fig. 2.1   The two different possibilities for the change δG in the free energy associated with 1st order and 2nd order phase transitions.

properties such as the specific heat $(C_p = T\partial S/\partial T = -T\partial^2 G/\partial T^2)$, the compressibility and the magnetic susceptibility of the system.

Let us now consider the effect on the free energy G of changing an external parameter, for instance the temperature. Such a change cannot introduce a sudden change in the energy of the system, because of the conservation of energy. Hence, if we consider the free energy per unit volume, g, of a system with a fixed number of particles and write $G = gV$, there are only two possibilities. Either the change δG in G arises from a change in the free energy density g, $\delta G = V\delta g$, or it comes from a change in the volume V, $\delta G = g\delta V$. When the properties of a system change as a result of a phase transition, they can undergo a small change δg all over the system at once or initially only in some parts δV of it, as shown in Fig. 2.1. If the new phase appears as $\delta G = g\delta V$, so that it appears only in parts δV of the system, then it requires the formation of stable nuclei, namely of regions of the new phase large enough for them to grow rather than to shrink. Since the energy consists of a negative volume term and a positive surface one, which for a spherical nucleus of radius r are proportional to r^3 and r^2 respectively, this critical size r_c is that for which the volume term equals the surface one, so that for $r > r_c$ growth of the nucleus leads to a decrease in its energy. Because of the need for nucleation, the first phase can coexist with the second phase, in a metastable state, even beyond the critical temperature for the phase transition. This is a first order phase transition. The best known manifestations of such a transition are superheating and supercooling.

In the other case, where the phase transition occurs simultaneously throughout the system, $\delta G = V \delta g$. Although the difference δg between the properties of these phases is small, the old phase which occupied the whole volume cannot exist, even as a metastable state, on the other side of the critical point, and it is replaced there by a new phase. These two phases are associated with different symmetries. For instance, in the paramagnetic state of a magnetic system there is no preferred direction, while in the ferromagnetic state there is a preferred direction, that of the total magnetic moment. In this case, the critical point is the end-point of the two phases, and so there must be some sudden change there, i.e., some discontinuity in their properties. This is an example of a second order phase transition.

Phase transitions are classified, as proposed by Ehrenfest, by the order of the derivative of the free energy which becomes discontinuous (or, in modern terms, exhibits a singularity) at the phase transition temperature. In a first order phase transition, a first derivative becomes discontinuous. A common example of this is the transition from a liquid to a gas when water boils, where the density shows a discontinuity. In a second order phase transition, on the other hand, properties such as the density or magnetic moment of the system are continuous, but their derivatives (which correspond to the second derivatives of the free energy), such as the compressibility or the magnetic susceptibility, are discontinuous. In this book, we will be concerned mainly with second order phase transitions, with which are associated many unusual properties.

2.2 Appearance of a Second Order Phase Transition

Before proceeding to a detailed mathematical analysis, it is worthwhile to consider qualitatively an example of how a second order phase transition can occur. Accordingly, we will now discuss the mean field theory of the paramagnetic–ferromagnetic phase transition in magnetic materials, originally proposed by Pierre Weiss some 100 years ago, in 1907 [2]. This consists of the sudden ordering of the magnetic moments in a system as the temperature is lowered to below a critical temperature T_c. He suggested that these materials consist of particles each of which has a magnetic dipole moment μ. For N such particles, the maximum possible magnetic moment of the system is $M_0 = N\mu$, when the moments of all the particles are aligned. Such a state is possible at $T = 0\,\text{K}$, when there is no thermal energy to disturb the orientation of the moments. In the presence of a small magnetic field H, the energy of a dipole of moment μ is $-\mu H$. For the sake of simplicity, we consider only two possible orientations of the dipoles, parallel and anti-parallel to the field, or up and down, and denote by N_+ and N_- respectively the number of dipoles in these two orientations at any given temperature. Similar results can be obtained if one allows the moments to adopt arbitrary orientations with respect to the field, but the analysis is slightly more complicated. Then the total magnetic moment of the system in the direction of the field is $M = (N_+ - N_-)\mu$, and its energy is $E = -MH$. The main assumption of Weiss was that there is some internal magnetic field acting on each of the dipoles, and that this field is proportional to M/M_0. This is a very reasonable assumption if the internal field on a given particle is due to the magnetic moments of the surrounding particles. Of course, it is an approximation to assume that each particle experiences the same magnetic field, and we will consider more refined theories later. We therefore write the effective field acting on each dipole (in the absence of an external field) in the form $H_m = CM/M_0$. According to Boltzmann's law, which was already known by then, the number of particles $N_\pm$ with moments pointing up and down at temperature T is proportional to $\exp(\mp\mu H_m/kT)$. Here and throughout the book, we denote the Boltzmann constant by k. It readily follows that

$$\frac{M}{M_0} = \frac{N_+ - N_-}{N_+ + N_-} = \tanh\left(\frac{T_c}{T}\frac{M}{M_0}\right) \equiv f\left(\frac{M}{M_0}\right), \qquad (2.2)$$

where $T_c = C\mu/k$. As can be seen from Fig. 2.2, if $T_c/T < 1$ then this equation only has the trivial solution $M = 0$, since for small arguments $\tanh(x) \simeq x$, and so there is no spontaneous magnetic moment if $T > T_c$. On the other hand, if $T < T_c$ then the equation has two solutions. An examination of the effect of a small change in the internal field shows that the solution with $M > 0$ is the stable one, i.e., the system has a spontaneous magnetic moment and so is ferromagnetic. The critical temperature T_c at which this transition from paramagnetism to ferromagnetism takes place is given by $T_c = C\mu/k$, the famous Weiss equation. At the time that Weiss wrote his paper, quantum mechanics, electronic spin and exchange effects had not been discovered, and so he could only estimate the strength of the internal field from the known dipole–dipole interactions, and this led to an estimate of T_c of around 1 K. Weiss knew that for iron T_c is around 1000 K, and so wrote bravely at the end of his paper [2] that his theory does not agree with experiments but future research will have to explain this discrepancy of three orders of magnitude. In spite of this discrepancy, his paper was accepted for publication, and we now know that his ideas on the nature of the paramagetic–ferromagnetic phase transitions are qualitatively correct.

2.3   Critical Phenomena

Phase transitions occur in Nature in a great variety of systems and under a very wide range of conditions. The thermodynamic functions become singular at the phase transition, and these mathematical singularities lead to many unusual properties of the system which are called "critical phenomena". We first consider the different types of the phase transition points ("critical points") and then we introduce

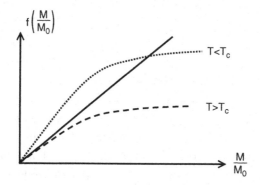

Fig. 2.2   Graphical solution of Eq. (2.2) for the magnetization M in the Weiss mean field model.

a qualitative method for describing the behavior of various parameters of the system in the vicinity of critical points.

The liquid–gas critical point of a one-component fluid is determined by the conditions [1]

$$\left(\frac{\partial p}{\partial v}\right)_T = \left(\frac{\partial^2 p}{\partial v^2}\right)_T = 0, \tag{2.3}$$

where p is the pressure, v is the specific volume, and T is the temperature. Similarly, the liquid–gas critical points of binary mixtures are characterized by the vanishing of the first and second derivatives of the chemical potential μ with respect to the concentration x,

$$\left(\frac{\partial \mu}{\partial x}\right)_{T,\,p} = \left(\frac{\partial^2 \mu}{\partial x^2}\right)_{T,\,p} = 0. \tag{2.4}$$

Here $\mu = \mu_1/m_1 - \mu_2/m_2$, where μ_1, μ_2 and m_1, m_2 are the chemical potentials and masses of the two components.

The close relation between (2.3) and (2.4) is evident from the equivalent form of Eq. (2.4), which can be rewritten as

$$\left(\frac{\partial p}{\partial v}\right)_{T,\,\mu} = \left(\frac{\partial^2 p}{\partial v^2}\right)_{T,\,\mu} = 0. \tag{2.5}$$

The critical conditions for a binary mixture (2.5) are the same as those for a pure system (2.3) when the chemical potential is kept constant. Analogously, the critical points for a n-component mixture are determined by the conditions

$$\left(\frac{\partial p}{\partial v}\right)_{T,\,\mu_1...\mu_{n-1}} = \left(\frac{\partial^2 p}{\partial v^2}\right)_{T,\,\mu_1...\mu_{n-1}} = 0, \tag{2.6}$$

where $n - 1$ chemical potentials are held constant.

In addition to the above-mentioned thermodynamic peculiarities, relaxation processes slow down near the critical points resulting in singularities in the kinetic coefficients. An example is the slowing down of diffusion near the critical points of a binary mixture. Nothing happens to the motion of the separate molecules when one approaches the critical point. It is the rate of equalization of the concentration gradients by diffusion which is reduced near the critical points. In fact, the excess concentration δx in some part of a system does not produce diffusion by itself. Usually a system has no difficulty in "translating" the change in concentration into a change in the chemical potential $\delta \mu \sim (\partial \mu / \partial x) \, \delta x$ which is the driving force for diffusion. However, near the critical point, according to (2.4), $\partial \mu / \partial x$ is very small, and the system becomes indifferent to changes in concentration. This is the simple physical explanation of the slowing down of diffusion near the critical point.

Since the states of a one-component system and a binary mixture are defined by the equations of state $p = p(T, \mu)$ and $\mu = \mu(T, p, x)$ respectively, Eqs. (2.3) and (2.4) define the isolated critical point for a one-component system, and the line of critical points for a binary mixture. Another distinction between one-component systems and binary mixtures is that there are two types of critical points in the latter: the above considered liquid–gas critical points and liquid–liquid critical points, whereas two coexisting liquid phases are distinguished by different concentrations of the components. Both critical lines are defined by Eq. (2.4).

Different binary liquid mixtures show either concave-down or concave-up coexistence curves in a temperature–concentration phase diagram (at fixed pressure) or in a pressure–concentration phase diagram (at fixed temperature). The consolute point is an extremum in the phase diagram where the homogeneous liquid mixture first begins to separate into two immiscible liquid layers. For the concave-down diagram, as for a methanol–heptane mixture, the minimum temperature above which the two liquids are miscible in all proportions is called the upper critical solution temperature (UCST). By contrast, for a concave-up diagram, such as the water–triethylamine solution, the maximum temperature below which the liquids are miscible in all proportions is called the lower critical solution temperature (LCST). Under the assumption of analyticity of the thermodynamic functions at the critical points, one can obtain the general thermodynamic criterion for the existence of UCST and LCST [3].

The properties of near-critical fluids range between those of gases and liquids (see Table 2.1). Near-critical fluids combine properties of gases and liquids. Their densities are lower than those of liquids, but much higher than the densities of gases, which makes the near-critical fluids excellent solvents for a variety of substances.

Table 2.1 Comparison of some physical properties of gases, liquids and near-critical fluids.

Physical Properties	Gas	Near-critical fluid	Liquid
Density (kg/m^3)	0.6–2	200–500	600–1000
Kinematic viscosity (10^{-6} m^2/sec)	5–500	0.02–0.1	0.1–5
Diffusion coefficient (10^{-6} m^2/sec)	10–40	0.07	$2 \times 10^{-4} - 2 \times 10^{-3}$

In Table 2.2 we list the critical parameters of the solvents in most common use. Water is the most abundant, cheap, safe and environmentally pure solvent. In spite of its high critical parameters which limits its application, in addition to the traditional uses, modern applications include the important problems of solving the environmental pollution problem and the fabrication of nanocrystalline materials with predictable properties [4]. Properties of near-critical water, such as the full mixing with oxygen and organic compounds, high diffusion and mass transfer coefficients, make water appropriate for efficient treatment of industrial wastes. The use of near-critical water for detoxification of organic waste using the catalytic oxidation of pyridine was found [5] to be cheaper than other methods and also more effective, having almost no limitation on the concentration of the pyridine-containing solutions. The efficiency of hydrothermal detoxification of pyridine waste is substantially increased by the addition of a small amount of heterogeneous catalyst. For instance, the addition of 0.5% of $PtAl_2O_3$ increases the oxidation of pyridine to 99% [6]. Other methods include dechlorination of chlorinate organic compounds, cleaning of polymers and plastic wastes, hydrolysis of cellulose, and the release of bromine for polymers and plastic wastes, hydrolysis of cellulose, and the release of bromine for polymers and plastics.

Nanocrystallines (particles whose sizes are a few interatomic distances) are new generation materials widely used as sensors, fuel cells, high-density ceramics, and semiconductors, among others. Hydrothermal synthesis in near-critical water is used to obtain nanocrystalline oxide powders with specified particle sizes and phase

Table 2.2 Critical parameters of fluids, which are commonly used as solvents for chemical reactions.

Solvent	T_{cr} (C)	p_{cr} (atm)	ρ_{cr} (g/mL)
Water (H_2O)	373.9	220.6	0.322
Carbon dioxide (CO_2)	30.9	72.9	0.47
Sulfur hexafluoride (SF_6)	45.5	36.7	0.73
Ammonia (NH_3)	132.3	113.5	0.235
Methanol (CH_3OH)	239.4	80.9	0.272
Propane (C_3H_6)	96.6	41.9	0.22
Ethane (C_2H_6)	32.2	48.2	0.20
Pyridine (C_5H_5N)	347	55.6	0.31
Benzene (C_6H_6)	289	48.3	0.30

composition. Many references can be found [4] dealing with both nanotechnology and environmental problems.

Like water, carbon dioxide (CO_2) has the advantage of being non-flammable, nontoxic and environment compatible. At the same time CO_2 has critical parameters more convenient than water, and is, therefore, the first choice for use as a near-critical solvent. Another advantage of CO_2 lies in the fact that it does not attack enzymes and is suitable for enzyme-catalyzed reactions. Some new applications include the use of two-phase reaction mixtures with high pressure carbon dioxide which are known as "CO_2-expanded fluids." In fact, near-critical carbon dioxide is more frequently used in the laboratory and in technology than any other solvent. Hundreds of examples can be found in recent reviews [7]–[8].

2.4 Correlations

For a second order phase transition, a second derivative of the free energy diverges as the phase transition is approached. For instance, the magnetic susceptibility $\partial M/\partial H = -\partial^2 G/\partial H^2$ tends to infinity as $T \to T_c$. Now according to the fluctuation–dissipation theorem [1], the magnetic susceptibility is proportional to the integral over all space of the average of the product of the magnetic moment at two points distance r apart, which describes the correlation between the magnetic moments at these points,

$$\frac{\partial M}{\partial H} \sim \int \langle M(0)M(r)\rangle \, d\tau \, . \tag{2.7}$$

In general, the magnetic moment (spin) at any site tends to align the spin at an adjacent site in the same direction as itself, so as to lower the energy. However, this tendency is opposed by that of the entropy, so that far from the critical point there is a finite correlation length ξ such that

$$\langle M(0)M(r)\rangle \sim \exp(-r/\xi) \, . \tag{2.8}$$

Here, the correlation length ξ has the following physical meaning. If one forces a particular spin to be aligned in some specified direction, the correlation length measures how far away from that spin the other spins tend to be aligned in this direction. In the disordered state, the spin at a given point is influenced mainly by the nearly random spins on the adjacent points, so that the correlation length is very small. As the phase transition is approached, the system has to become "prepared" for a fully ordered state, and so the "order" must extend to larger and larger distances, i.e., ξ has to grow. However, the divergence of the integral in Eq. (2.7) as $T \to T_c$ implies that at the critical point the correlation function cannot decrease exponentially with distance r, but rather must decay at best as an inverse power of r,

$$\langle M(0)M(r)\rangle \sim r^{-\gamma}, \quad \gamma \le 3 \, . \tag{2.9}$$

This is a point of great physical significance. It means that near the critical point not only do we not have any small energy parameter, since the critical temperature is of the same order of magnitude as the interaction energy, but also we do not have any typical length scale since the correlation length diverges on approaching the critical point. In other words, all characteristic lengths are equally important near the critical point, which makes this problem extremely complicated. A similar situation of various characteristic lengths arises in the problem of the motion of water in an ocean, but these are associated with different phenomena. The angstrom–micron length scale is appropriate for studying the interactions between water molecules, but one must take into account lengths on the order of meters for studying the tides and the kilometer length scale for studying the ocean streams. This is in contrast to the situation near critical points, where one cannot perform such a separation of different length scales.

We now return to the question raised at the beginning of this chapter, namely how one can obtain long-range correlations from short-range interactions. In mathematical terms, the question is how an exponentially decaying correlation can transfer the mutual influence of different atoms located far away from each other. A qualitative answer to this question has been given by Stanley [9]. The correlations between two particles far apart do indeed decay exponentially. However, the number of paths between these two particles along which the correlations occur increase exponentially. The exponents of these two exponential functions, one positive and one negative, compensate each other at the critical point, and this leads to the long-range power law correlations. By contrast, for a one-dimensional system the exponential increase corresponding to the number of different paths is replaced by unity, and so the negative exponent leads to the absence of ordering and so to no phase transition for non-zero temperatures.

Curiously enough, in the Red Army of the former Soviet Union the order given by an officer standing in front of the line of soldiers was "Attention! Look at the chest of the fourth man!". For some unclear reason, they decided that the correlation length is equal to four, and the soldiers will be ordered in a straight line if each one is aligned with his fourth neighbor in the row.

2.5   Speed of Equilibration Near the Gas–Liquid Critical Point

Due to the divergence of the isothermal compressibility and the isobaric thermal expansion, as well as the vanishing of thermal conductivity and sound velocity at the critical point, the characteristic property of all processes near the critical point is slowing down, which is the main problem for experimentalists. However, after performing some spacelab calorimetric experiments, it become clear that the opposite process occurs — the speeding up of the equilibration processes. The latter is connected with replacing the slow diffusion process by the thermo-acoustic propagation of the boundary layer inside a system, which leads to thermalization and homogeneity of the system. The theoretical explanation of this interesting phenomenon is based on the piston effect, which is due to the macroscopic manifestation of critical fluctuations and leads to the appearance of the viscous dissipation function Φ in the diffusion equation, which is responsible for the speed of the process. Indeed, for a liquid at rest ($v = 0$),

$$\rho T \frac{\partial s}{\partial t} = div \left(\lambda \bigtriangledown T \right) , \tag{2.10}$$

where s is the entropy per unit mass and λ is the thermal conductivity. For an incompressible liquid, changes in the entropy are only due to the temperature variation, whereas near the critical point, the compressibility is very high and one has to take into account the change of the pressure [10],

$$\frac{\partial s}{\partial t} = \left(\frac{\partial s}{\partial T} \right)_p \frac{\partial T}{\partial t} + \left(\frac{\partial s}{\partial p} \right)_T \frac{\partial p}{\partial t} . \tag{2.11}$$

Combining equations (2.10) and (2.11) and performing simple thermodynamic transformations yields

$$\frac{\partial T}{\partial t} = \frac{1}{\rho c_p} div \left(\lambda \bigtriangledown T \right) + \Phi , \tag{2.12}$$

where

$$\Phi = \left(1 - \frac{c_v}{c_p} \right) \left(\frac{\partial T}{\partial p} \right)_\rho \frac{\partial p}{\partial t} . \tag{2.13}$$

Integration of Eq. (2.11) over the fixed volume of the system gives

$$\frac{\partial p}{\partial t} = - \int (\partial \rho / \partial T)_P \left(\partial T / \partial t \right) d \left(1/\rho \right) \left[\int (\partial \rho / \partial p)_T \, d \left(1/\rho \right) \right]^{-1} , \tag{2.14}$$

where it was assumed, as in (2.13), that due to the small size of the system, $\partial p / \partial t$, like p, is independent of the spatial variables.

Numerical calculations for xenon show [10] that the second term in Eq. (2.12) is much larger that the first term, demonstrating thereby that the transient thermal response to temperature changes on the boundary is much faster than the thermodiffusial process. It turns out that the thermal relaxation was completed within seconds, rather than the days predicted by the thermal diffusion model. The crucial role of the transient thermal response in the process of thermalization was experimentally proved in the critical regions of SF_6 [11] and CO_2 [12], [13]. In our analysis, we neglected gravity, which becomes very important in experiments on critical fluids. The analysis of the effect of gravity on heat transfer has been performed [14], which shows that temperature stabilization has not changed, but stratification of density under gravity takes hours, rather than seconds, which is however still faster than the time required for thermal diffusion.

2.6 The Mechanism of Slowing Down Near the Critical Points

The slowing down of the dynamic processes near the liquid–gas critical points is the hallmark of the critical phenomena. Although the very interesting phenomena considered in the previous section of speeding up of a thermal equilibrium established through the thermoacoustic effect, where the temperature change at the surface induces acoustic waves which result in a fast change of the pressure ("piston effect"), and, hence, of the temperature everywhere in the fluid, the slowing down is still the one which makes it so difficult to get reliable equilibrium experimental data near the critical points. Therefore, the elucidation of the mechanism of the slowing down is of a great importance. This is just the goal of our studies.

In order to explain what we mean by "mechanism", let us recall another characteristic of phase transitions mentioned in section 2.3, the slowing down of diffusion near the critical point of binary mixtures, the peculiar phenomenon which is well established experimentally. These critical points are characterized by the vanishing of the derivative of the chemical potential μ with respect to the concentration x of one of the component, $\partial\mu/\partial x = 0$. When the concentration is increased at some region in the liquid, one expects that the diffusion of this component from the region will make a system more homogeneous. However, the driving force for diffusion is the gradient of the chemical potential $\nabla\mu$, and not ∇x, and since in the near critical region $\partial\mu/\partial x$ is very small, a system is unable "transform" ∇x into $\nabla\mu = (\partial\mu/\partial x)\nabla x$. This effect is purely thermodynamic, and one concludes that the "mechanism" of the slowing down of diffusion is purely phenomenological (macroscopic) wherein nothing happened with the (self-) diffusion of individual molecules, they don't "feel" the approach to a critical point. The diffusion is slowing since, in spite of the excess concentration somewhere in a system, on the average the same number of molecules cross each plane in two opposite directions.

In our case of the Ising model, the spin of a particle changes its direction from time to time due to the thermal fluctuations. The question arises whether such jumps become slower when the critical point approaches, which would explain the slowing down of the dynamic processes near the critical point, or, like in the case of binary mixture, there is another factor responsible for the slowing down. In fact, as we show by the numerical calculations such an another factor exists indeed. It turns out that among many consecutive random jumps of each given spin, those which bring the spin back to its original state occur more often, which means that no changes occur, the diffusion is slowing. The latter is just the "mechanism" of the slowing down rather than intuitive assumption of the slowing down of the individual jumps.

2.7 Conclusion

Phase transitions are very general phenomena which occur in a great variety of systems under very different conditions. They can be divided into first- and second-order transitions depending on which derivatives of the free energies have anomalies at the transition. The existence of phase transitions, as such, was established a hundred years ago in the framework of the mean field theory. Three major factors which present severe difficulties for the theoretical description of phase transitions are the non-analyticity of the thermodynamic potentials, the absence of small parameters, and the equal importance of all length scales.

Chapter 3

The Ising Model

We now consider a microscopic approach to phase transitions, in contrast to the phenomenological approach used in the previous chapter. In this approach, following Gibbs, we start with the interaction between particles. The first step is then to calculate the mechanical energy of the system E_n in each state n of all the particles, a problem which in general is far from trivial for a system of 10^{23} particles. In the framework of classical and quantum mechanics we must then calculate the partition function

$$Z = \sum_n \exp\left(-\frac{E_n}{kT}\right), \qquad Z = Tr\left[\exp\left(-\frac{H}{kT}\right)\right], \qquad (3.1)$$

respectively, where H is the system Hamiltonian, and the Helmholtz free energy F of the system is

$$F = -kT \ln Z. \qquad (3.2)$$

For a large enough system, one can replace the summation over n in Eq. (3.1) by an integration over phase space, so that for non-interacting particles the integral over the coordinates of all the N particles equals V^N. It follows that $F = -NkT \ln(CV)$, where C is independent of V. On using Eq. (2.1), we find that

$$P = -\left(\frac{\partial F}{\partial V}\right)_T = NkT/V, \qquad (3.3)$$

which is just the equation of state of an ideal gas. This equation involves only two degrees of freedom, instead of the around 10^{23}

degrees of freedom of the original system. In this way, one can proceed from a detailed microscopic description of the system to a simple thermodynamic description. However, this method only applies to systems in equilibrium, while for non-equilibrium systems there is no unique approach, a point that will be considered in Chapter 12. Even for systems in equilibrium this description is only simple in principle, but not in practice since it involves first calculating the mechanical energy of the numerous different possible many-particle states and then a summation (or integration for continuous variables) over all the possible states. Only for some special simple systems it is possible to perform the calculations exactly, but it is very instructive to do so for such systems and examine the results that are obtained.

Let us mention that before the seminal work of Onsager [15], it was not at all clear whether statistical mechanics is able to describe the phenomena of phase transitions, i.e., how the "innocent" expression involving T in Eq. (3.1) will lead to non-trivial singularities at some specific temperature. The answer lies in the fact that the singularities appear only for a system of infinite size (the thermodynamic limit) which has an infinite number of configurations. It is just the infinite number of terms in the sum which appears in Eq. (3.1) that can lead to singularities.

One of the simplest model systems is the so-called Ising model, which we will now examine. This model, which will be discussed extensively in this book, is based on the following three assumptions:

(1) The objects (which we call particles) are located on the sites of a crystal lattice.

(2) Each particle i can be in one of two possible states, which we call the particle's spin S_i, and we choose $S_i = \pm 1/2$.

(3) The energy of the system is given by

$$E = -J \sum_{i,\,j} S_i S_j \,, \tag{3.4}$$

where J is a constant and the sum is over all pairs of adjacent particles, i.e., over all pairs of nearest neighbors i and j. Thus, in a linear lattice each particle interacts only with its two nearest neighbors, in a square lattice with its four nearest neighbors, and in a simple cubic lattice with its six nearest neighbors.

In spite of its simplicity, the Ising model is used in many applications where an object can be in one or two states, such as sites occupied by A or B atoms, sites containing a particle or a hole, two possible conformations, and even votes for one of two political parties in elections. One can generalize the first two basic assumptions of the Ising model. A larger number of possible states is taken into account in the so-called Potts model [16], and one can consider not only interactions between nearest neighbors but also interactions between second nearest neighbors, third nearest neighbors, etc. These generalizations make the problem much more complicated. On the other hand, the lattice approximation seems to be of no importance for phase transition problems since, as we will see later, much longer distances (on the order of the correlation length) are important in phase transitions.

The personal story of Ising is also of interest [17]. In 1924–1926, Ising was a doctoral student of the famous German physicist Wilhelm Lenz, who suggested this model to him. In his thesis, Ising showed that the system does not exhibit a phase transition in one dimension, which is correct. He also showed that there is no phase transition in two dimensions, which was shown much later (by Onsager in 1944) to be incorrect. After completing his thesis he started to work as a high-school teacher, and with the rise to power of the Nazis he went to Luxembourg. We next hear of him in 1948 in the USA, where he taught at some small university. He died in 1990, and during his whole career published only two scientific papers, one on his thesis work and the other entitled "Goethe as a physicist". However, on his arrival in America in 1948 he found big placards announcing a conference on the Ising model, and this model is still being extensively studied as a paradigm of a simple system exhibiting a well-defined phase transition.

3.1   Phase Transitions in the Ising Model

For the Ising model with spins $S = \pm 1/2$ pointing up or down, the exchange energy is

$$E = \sum_{i,j} J_{ij} S_i S_j \tag{3.5}$$

where the summation is over nearest neighbors i and j. Replacing S_i by $r_i = (1 - 2S_i)/2$, leads to r_i equal to zero or one, which describes the lattice gas model, where at each site there can be one ($r_i = 1$) or zero ($r_1 = 0$) particles. These models are also isomorphic to the binary $A - B$ mixture, where $n_i(A) = 1$ and $n_i(B) = 0$ describes particle A or B located at site i so that

$$n_i(A) + n_i(B) = 1. \tag{3.6}$$

Consider two nearest neighboring sites, i and j. They can be occupied by two atoms A or by two atoms B with the interaction energies ε_{AA} and ε_{BB}, respectively. Another possibility is the configurations AB or BA with energy ε_{AB}. Therefore, the energy of the $A - B$ mixture will be

$$E = \sum_{i,j} \{ \varepsilon_{AA} \, n_i(A) n_j(A) + \varepsilon_{BB} \, n_i(B) n_j(B)$$

$$+ \, \varepsilon_{AB} \, [n_j(A) n_j(B) + \, n_i(B) n_j(A)] \}. \tag{3.7}$$

Inserting $n(B)$ from (3.6) to (3.7) yields

$$E = \sum_{i,j} [\varepsilon_{AA} + \varepsilon_{BB} - 2\varepsilon_{AB}] \, n_i(A) n_j(A) + O\{n(A)\}, \tag{3.8}$$

where $O\{n(A)\}$ contains terms that do not effect the interaction between two A particles. As it follows from the last equation, for $[\varepsilon_{AA} + \varepsilon_{BB} - 2\varepsilon_{AB}] < 0$, the configuration of the same atoms, AA or BB, becomes energetically more favorable and a disordered system will dissociate into two ordered A and B phases.

3.2   1D Ising Model

We consider first the one-dimensional (1D) Ising model, and examine how to calculate the partition function. Let us assume that the system contains N particles, for which the partition function has the following form

$$Z_N = \sum_{S_j = \pm 1/2} \exp\left[-\frac{J}{kT}(S_1 S_2 + S_2 S_3 + \cdots + S_{N-1} S_N) \right] . \quad (3.9)$$

For a system with $N + 1$ particles, the sum contains one additional term, so that

$$Z_{N+1} = Z_N \sum_{S_N, S_{N+1} = \pm 1/2} \exp\left(-\frac{J}{kT} S_N S_{N+1} \right) . \quad (3.10)$$

Since $S_N S_{N+1} = \pm 1/4$, it follows that $Z_{N+1} = 2\cosh(J/4kT)Z_N$, and so (since $Z_1 = 2$), by induction,

$$Z_N = 2^N \left(\cosh \frac{J}{4kT} \right)^{N-1} . \quad (3.11)$$

Hence if $N \gg 1$ the free energy is

$$G = -kT \ln Z = -NkT \ln \left(2\cosh \frac{J}{4kT} \right) . \quad (3.12)$$

Since this is a monotonic function of T, with no singularity except at $T = 0$, the 1D Ising system cannot exhibit a phase transition at any finite temperature.

Another interesting general type of system is the one in which the interactions between the particles are very weak but of infinite range,

$$\varphi(r) = -\lim_{\gamma \to 0} \gamma \exp(-\gamma r) , \quad (3.13)$$

in contrast to the strong short-range interactions of the Ising model. For this model there is a phase transition even in one dimension, and we will discuss it in Chapter 8. In fact, it is not necessary that there should be interactions between all the spins in a system for a phase transition to occur. Even a few random long-range interactions combined with short-range interactions, as in the Ising model, are sufficient to produce a phase transition, as in the small world model which we discuss in Chapter 12.

As we have seen, due to the predominance of the entropy (disordered) factor compared to the (ordered) energy factor, there is no disorder–order phase transition in a one-dimensional equilibrium system with short-range interactions. However, a small modification of such a model leads to phase transitions, which have different applications. One of these applications is the thermal melting transition (denaturation or unzipping) of DNA. At some phase transition temperature or at some pH of the system, the doubled-stranded DNA unzips into two separate coils. The simplest approach is the molecular zipper model proposed by Kittel as a simple model of KH_2PO_4 [19]. The model consists of a double-ended zipper (N parallel links) that can be opened from only one side. If the first p links are open, the energy required to open the $p + 1$ link is e. However, if all the preceding links are not open, this energy is infinite. The last $p = N$ link is always closed, i.e., the zipper is open when $N - 1$ links are open. The open link may have one of G different orientations associated with the rotation freedom of a link. A simple calculation of the statistical sum [19] or the transfer matrix [20] shows that such a system has a first order phase transition

Another one-dimensional model with short range interactions that allows a phase transition was proposed by Chui and Weeks [21]. In simplified form, their model is described by the Hamiltonian in a one-dimensional lattice with periodic boundary conditions of the form [23]

$$H = \sum_{i=1}^{N} \left\{ J \mid n_{i+1} - n_i \mid -B\delta\left(n = 0\right) \right\}, \qquad (3.14)$$

where n is an integer, n_i is the distance between the substrate and interface position at site i, and N is the total number of sites. The first term in Eq. (3.14) is a sort of surface tension, and the second term is a potential energy of strength B, which binds the interface to the substrate at the site $n_i = 0$. Chui and Weeks show analytically [21] that this model represents a thermodynamic one-dimensional phase transition at temperature $T = T_C$, where $1 - \exp\left(-J/\kappa T_C\right) = \exp\left(-B/\kappa T_C\right)$, which separates the bound ordered states at low temperatures $T < T_C$ from the disordered states at high temperatures $T > T_C$. The more sophisticated Peyrard–Bishop–Dauxois model [22] contains one degree of freedom y_n for each base pair, describing their separation from the ground state position. Due to the more detailed description of the interaction between base pairs, this model is able to describe not only the phase transition, but also the bubble nucleation dynamics in the pre-transition regime as well as the statistical and dynamic behavior of long chains. The model is defined by the following Hamiltonian,

$$H = \sum_{i=1}^{N} \left\{ \frac{m}{2} \left(\frac{dy}{dt}\right)^2 + D\left[\exp\left(-a_i y_i\right) - 1\right]^2 + W\left(y_i, y_{i-1}\right) \right\},$$
(3.15)

where the first term is the kinetic energy for bases of mass m and the second term is the Morse potential, which describes the interaction between opposite bases. The last term has the following form,

$$W = \frac{k}{2} \left\{1 + \rho \exp\left[-\alpha\left(y_n - y_{n-1} - 1\right)\right]\right\} \left(y_n - y_{n-1}\right)^2,$$
(3.16)

which describes the nearest-neighbor stacking interaction between base pairs along each chain. In addition to the last harmonic factor in Eq. (3.16), this equation contains the stiffness parameter ρ, which is responsible for the different effective coupling constant along the chains. Numerical thermodynamic calculations of this model yield good agreement with experiment. In particular, below some critical temperature, the chains remain bound, whereas above this temperature, where the mean value of y_i diverges, the chains separate.

3.3 2D Ising Model

The exact solution of the 2D Ising model was presented by Onsager in 1944 in a very long and complicated paper [15], and even after a few simplifications, a recent version of the proof took seven pages in the very concisely written classic text of Landau and Lifshitz [1]. We

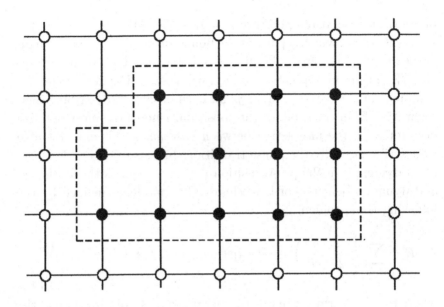

Fig. 3.1   2D Ising square lattice, with an island of opposite spins (indicated by black solid dots) of perimeter L. (B. Liu and M. Gitterman, *Am. J. Phys.* **71** 806 (2003). Copyright 2003 by the American Physical Society.)

will now present some qualitative arguments for analyzing the phase transitions in the 2D Ising model.

Since we want to analyze the role of the energy E and the entropy S in order–disorder phase transitions, it is convenient to consider the Helmholz free energy $F = E - TS$. At all temperatures the stable state corresponds to the minimum value of F. At low temperatures the energy E plays the leading role, and its minima correspond to ordered states. At high temperatures the entropy S dominates F, and so the minima of F are reached when the entropy S is a maximum. Therefore, at some intermediate temperature $T = T_c$ the ordering influence of the energy and disordering influence of the entropy are balanced, and one can qualitatively estimate the critical temperature T_c as

$$T_c \approx \frac{\Delta E}{\Delta S}. \tag{3.17}$$

In order to apply this equation to the 2D Ising model, we consider the two competing states of the Ising lattice shown in Fig. 3.1 [18]. The question arises as to whether the fully ordered state 1 is able to transform spontaneously into state 2, which contains an island of opposite spins with perimeter of length L. For this to happen, the free energy of state 2 must be lower than that of state 1. To examine when this occurs, let us consider the changes in energy and in entropy when such an island of opposite spins is formed. Since at each site on the perimeter of the island the energy is raised by $2J$, the energy required to form the island is $\Delta E = (2J)L$. On the other hand, the number W of microscopic possibilities for creating such an island is approximately 3^L, since from each site one can continue the perimeter in three different directions, and so the change in entropy is $\Delta S \approx \ln(3^L N)$, where the factor N arises from an approximate evaluation of the number of ways in which the initial point can be chosen. This estimate is also approximate because we neglect the necessity to come back to the initial site after L steps, and also the requirement that the perimeter cannot cross itself. In this approximation, the change in the free energy of the system is

$$\Delta F = \Delta E - T\Delta S = L[2J - kT \ln 3] - kT \ln N \qquad (3.18)$$

and for L of order N the term $\ln N$ in this equation can be neglected. Therefore, a disordered state is favored if $T > T_c$, where $\kappa T_C = 2J/\ln 3$. This is not a bad approximation to the exact result found by Onsager [15],

$$kT_c = 2J/\ln(1 + \sqrt{2}) \qquad (3.19)$$

and our analysis shows clearly how the phase transition arises as a result of the competition between energy and entropy. Incidentally, for the 1D case instead of an island of perimeter L one has a change in the direction of the spin at just one point, so that $\Delta E = 2J$. On the other hand, since this point can be anywhere, $W = N$, so that

$$\Delta F = \Delta E - T\Delta S = 2J - kT \ln N \qquad (3.20)$$

which only becomes negative when $kT_c = 2J/\ln N$, so that $T_c \to 0$ as $N \to \infty$. This is in agreement with our previous conclusion that for an infinite 1D system, a phase transition can only occur at $T = 0$.

There is an important general conclusion from our above analysis of the 2D Ising model. Both our calculations and the exact one of Onsager show that kT_c is close to J, which is the only energy parameter appearing in our problem. Thus, there is no small parameter, which is not a happy situation for a physicist, since the theory of many-body problems usually relies on expressing properties as a power series in a small parameter, such as the ratio of the average potential energy to the average kinetic energy per particle for a weakly non-ideal gas, and the opposite ratio for solids. This is exactly the reason for the absence of a general rigorous theory of liquids, where these two average energies are of the same order of magnitude. It is also one reason why it took so many years to make progress in solving the problem of phase transitions.

Following the little-known article of Svrakic [24], we now present a very simple argument to "derive" exactly the result of Onsager for the 2D Ising model. Instead of considering the whole system and islands of arbitrary size, let us consider only an elementary cell containing just 4 sites. Since on each site the spin can be up or down, there are only $4^2 = 16$ possible spin configurations in such a cell, as shown in Fig. 3.2, and the energies of each can readily be calculated. As can be seen, there are two types of "ordered" states, with energies $\pm 4J$, each of which is doubly degenerate (spins up or down), and twelve "disordered" states with energy zero. The names "ordered" and "disordered" are connected with the fact that by putting together either of the first two "ordered" configurations one can construct an isotropic infinite two-dimensional Ising lattice, while this cannot be done for the "disordered" configurations. In accordance with the idea of competition between ordered and disordered states, we conjecture that the phase transition occurs when the partition function of the ordered states,

$$2 \exp\left(\frac{4J}{kT}\right) + 2 \exp\left(\frac{-4J}{kT}\right) \tag{3.21}$$

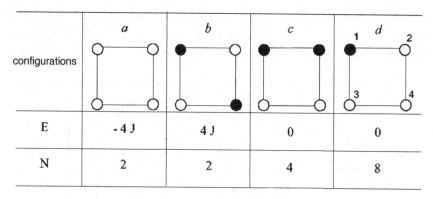

Fig. 3.2   Different configurations of an elementary cell of a 2D Ising square lattice divided into four groups, when the energy E and number of configurations N are shown for each group. Groups a and b contain non-degenerate configurations while degeneracies are inherent in groups c and d. (B. Liu and M. Gitterman, *Am. J. Phys.* **71** 806 (2003). Copyright 2003 by the American Physical Society.)

equals that of the disordered ones, 12, i.e., when $x^2 + x^{-2} = 6$, where $x = \exp(2J/kT)$. The solution of this equation is $x = 1 + \sqrt{2}$, so that $kT_c = 2J/\ln(1 + \sqrt{2})$, which quite surprisingly gives the exact result of Onsager. However, this method as presented above does not work for a three-dimensional Ising lattice.

3.4   3D Ising Model

It is interesting to apply the method described above to the Ising model for three-dimensional lattices, and to find the critical temperature of these lattices by considering just one elementary cell [25]. All the possible $2^8 = 256$ configurations of an elementary cell for a simple cubic lattice are shown in Fig. 3.3, where, as in Fig. 3.2 for the 2D lattice, N represents the number of equivalent configurations associated with the appropriate group. Let us consider first the different configurations of the 2D Ising lattice shown in Fig. 3.2. We call the configurations of groups a and b "non-degenerate" in the sense that all neighboring states, i.e., those that can be obtained from a given one by the flipping of a single spin, have an energy different from that of the original one. By contrast, all the configurations contained in group c do not change their energy if one of the spins is flipped. Hence, these configurations have adjacent states of the same

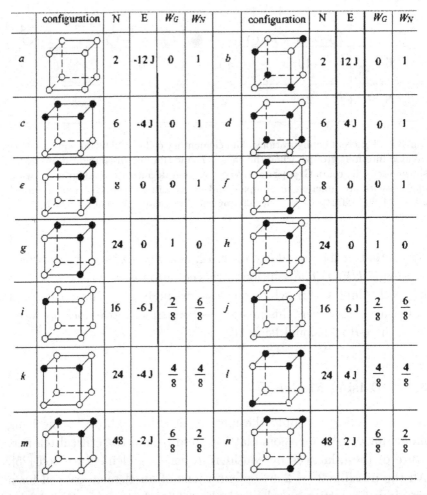

configuration	N	E	W_G	W_N	configuration	N	E	W_G	W_N
a	2	-12 J	0	1	b	2	12 J	0	1
c	6	-4 J	0	1	d	6	4 J	0	1
e	8	0	0	1	f	8	0	0	1
g	24	0	1	0	h	24	0	1	0
i	16	-6 J	$\frac{2}{8}$	$\frac{6}{8}$	j	16	6 J	$\frac{2}{8}$	$\frac{6}{8}$
k	24	-4 J	$\frac{4}{8}$	$\frac{4}{8}$	l	24	4 J	$\frac{4}{8}$	$\frac{4}{8}$
m	48	-2 J	$\frac{6}{8}$	$\frac{2}{8}$	n	48	2 J	$\frac{6}{8}$	$\frac{2}{8}$

Fig. 3.3   Different configurations of the elementary cell of a 3D Ising cubic lattice with N the number of equivalent configurations in each group, and E their energy. Groups a–f contain non-degenerate (stable) configurations, while the configurations in groups g and h are degenerate (unstable). Groups i–n make a contribution to the partition function of both degenerate and non-degenerate states, which are characterized by the fractional statistical weights W_G and W_N respectively. (B. Liu and M. Gitterman, *Am. J. Phys.* **71** 806 (2003). Copyright 2003 by the American Physical Society.)

energy, and with respect to this property we call these configurations "degenerate". The situation is slightly more complicated for the configurations belonging to group d. Here, flipping spin 1 will change the energy by $4J$, while flipping spin 4 will lead to a change in energy of $-4J$. Due to the symmetry of the latter two states with respect to the original one with $E = 0$, we place this "quasi-degenerate" state in the "degenerate" group. We note that flipping spins 2 or 3 will not change the energy. Hence, all configurations of group d are related, along with those of group c, to "degenerate" configurations. Our conjecture consists of the statement that the phase transition occurs when the partition function Z_{ndg} of non-degenerate states is equal to that Z_{dg} of degenerate states,

$$Z_{ndg} = Z_{dg} \qquad (3.22)$$

i.e., we identify the non-degenerate states with "ordered" configurations and the degenerate states with "disordered" ones.

Since for the two-dimensional Ising lattice our criterion (3.22) coincides with that of Svrakic [24], it clearly results in the exact Onsager solution for the critical temperature. However, in contrast to [24], our criterion (3.22) can be applied to three-dimensional lattices as well. We now examine Fig. 3.3 in detail. Here, groups a–f belong to the non-degenerate category, because a flip of the spin of any site will change the energy of the cell, and no quasi-degenerate states are observed in these configurations. Groups g and h belong to the degenerate category. A flip of each spin of group g results in the change of energy of $2J, -2J, 2J, -2J, 2J, -2J, 2J, -2J$, while a flip of each spin of group h change the energy by $6J, -6J, 6J, -6J, 2J, -2J, 2J, -2J$, and these two groups are totally quasi-degenerate. However, groups $i - n$ contribute to both degenerate and non-degenerate states. Let us consider, as an example, group i. The flip of each of the eight spins changes the energy by $-6J, 6J, 6J, 6J, 6J, 2J, 2J, 2J$, respectively. The first two excited states are symmetric, and, according to our classification, they are quasi-degenerate, while the remaining six states are non-degenerate. The numbers shown in Fig. 3.3 are the statistical weight of degenerate W_G and non-degenerate W_N states, and so for group i they have the fractional values $2/8$ and $6/8$ respectively. Analogous analyses can easily be performed for each of the groups $j-n$, and the final result for the solution of Eq. (3.22) is $kT_c/J = 4.277$ [25] which is much closer to the numerical result $kT_c/J = 4.511$ [26] than the value obtained by the method of Ref. [24], $kT_c/J = 2.030$.

In conclusion, we have shown that the critical temperature of two- and three-dimensional Ising models can be obtained by simple physical arguments based on the compromise between configurations which are "degenerate" or "non-degenerate" with respect to a single spin-flip excitation. A challenging problem for the interested reader is to expand our calculations to other elementary cells of two-dimensional lattices (say, 3–3 square, 3–2 rectangle, etc.) and to other types of three-dimensional lattices (BCC, FCC, etc.). For the latter, the findings have to be compared with the precise numerical results which are 6.235 and 9.792 for the BCC and FCC lattices, respectively [26]. The results of such a comparison are not known *a priori* since it is not clear how well this method will work in the general case.

3.5   Conclusion

The microscopical analysis of phase transitions can be performed only for simple models such as the Ising model. The results for these show that for non-zero temperature a phase transition does not appear in a one-dimensional system but such a transition does occur in two dimensions. A qualitative explanation of the onset of a phase transition as an order–disorder transition is that it is a result of competition between the ordering tendency of the energy and the disordering tendency of the entropy. With some simplified conjectures, such considerations can even lead to quantitatively correct results.

Chapter 4

Mean Field Theory

As we mentioned in Chapter 2, the simplest way to allow for interactions between particles is the mean field approximation, in which one assumes that the environment of each particle corresponds to the average state of the system and determines this average state self-consistently.

In order to understand the nature of Weiss' mean field approximation described in Chapter 2, it is useful to examine the application of this approximation to the Ising model. In this model, the energy of a state of the system with a given arrangement $\{S_1, S_2, \ldots, S_N\}$ of the N spins S_j, each of which can have the value $\pm 1/2$, situated on a lattice of sites, is

$$E = -J \sum_{i,\,j} S_i S_j \,. \tag{4.1}$$

In Weiss' mean field approximation for our system,

$$E_{mf} = -\sum_i S_i H_m \,, \tag{4.2}$$

where H_m is the effective magnetic field. A comparison of Eqs. (4.1) and (4.2) shows that $H_m = J\langle S_j \rangle$, so that

$$E_{mf} = -J \sum_{i,\,j} S_i \langle S_j \rangle \,. \tag{4.3}$$

Now, Eq. (4.1) can be written in the form

$$E = -J \sum_{i,\,j} S_i \langle S_j \rangle - J \sum_{i,\,j} S_i [S_j - \langle S_j \rangle]. \tag{4.4}$$

The first term in this equation is the expression that we derived for the mean field approximation, while the second term arises from the fluctuations of each spin from its mean value. Hence the mean field approximation is equivalent to ignoring fluctuations. It is obvious that as the number of dimensions of the system increases, and with it the number of nearest neighbors of each site, the importance of these fluctuations decreases, and they are negligible for a system of sufficiently high dimensions. As we will see later on, "sufficiently high" for the Ising systems means $d > 4$. This is the physical reason why for each system there is an upper critical dimension above which the mean field approximation is valid.

4.1   Landau Mean Field Theory

We now consider the thermodynamic approach to mean field theory proposed by Landau. He suggested [1] introducing an extra parameter η, which is called an order parameter, to distinguish the phase of the system, so that the Gibbs free energy G is now a function of three parameters, $G = G(P, T, \eta)$, where $\eta = 0$ in a disordered state found, as a rule, at high temperatures, $T > T_c$, and $\eta \neq 0$ for the ordered state, which then occurs for $T < T_c$. Since the free energy of the system in equilibrium is uniquely determined by P and T, this order parameter must be a function of them, $\eta = \eta(P, T)$. The first assumption of Landau's mean field theory is that an order parameter can be defined such that $\partial G / \partial \eta = 0$ and $\partial^2 G / \partial \eta^2 > 0$, which is the natural condition for minima of the free energy. His second assumption is based on the fact that at the phase transition $\eta = 0$. Therefore, close to the phase transition point he assumed that G can be expanded as a power series in η,

$$G(P, T, \eta) = G_0(P, T) + \alpha \eta + A \eta^2 + \beta \eta^3 + B \eta^4 + \cdots \tag{4.5}$$

where all the coefficients are functions of P and T. The first assumption, $\partial G/\partial \eta = 0$, can only be satisfied for all η if $\alpha = 0$. For a magnetic system in which the magnetic moment vector is the order parameter, symmetry arguments show that it is not possible to form a scalar from η^3, and so $\beta = 0$. Incidentally, for different systems the order parameter can have different numbers of components. For instance, in the liquid–gas phase transition the order parameter is the difference in densities between the two phases, which is a scalar. In superfluidity and superconductivity, it is the wave function ψ which has two components (its real and imaginary parts, or its amplitude and phase). For the anisotropic Heisenberg ferromagnet, in which M has three independent components, this will also be true of η.

Here we restrict our attention to systems in which the vector η has just one component, and for which close to the phase transition point the free energy can be written in the form

$$G(P,T,\eta) = G_0(P,T) + A\eta^2 + B\eta^4 . \tag{4.6}$$

For the sake of simplicity, we consider systems at a constant pressure equal to the critical one, $P = P_c$, and so omit P from the arguments. Then for arbitrary η the requirement that $\partial G/\partial \eta = 0$ means that $2A\eta + 4B\eta^3 = 0$, so that $\eta = 0$ for $T > T_c$, and there is an additional solution $\eta^2 = -A/(2B)$ for $T < T_c$. From the second requirement $\partial^2 G/\partial \eta^2 > 0$ it follows that for $T > T_c$, $A > 0$ while $A < 0$ for $T < T_c$. It follows that $A(T_c) = 0$, and the simplest form for A that satisfies these conditions is $A = a(T - T_c)$, with $a > 0$. One then finds that the order parameter

$$\eta \sim \sqrt{T_c - T} \tag{4.7}$$

and the free energy

$$G(T,\eta) = G_0(T) + a(T - T_c)\eta^2 + B\eta^4 . \tag{4.8}$$

Here the term in η^4 is required since the term in η^2 vanishes at the critical point, but the temperature dependence of its coefficient B can be ignored, so that we can write $B = B(T_c, P_c)$.

We now consider the implications of Eq. (4.8) for the free energy. From the formula for the entropy

$$S = -\frac{\partial G}{\partial T} = S_0(T) - a\eta^2 \tag{4.9}$$

it follows that

$$\begin{aligned} S &= S_0(T), & T &> T_c, \\ S &= S_0(T) + \frac{a^2(T - T_c)}{2B}, & T &< T_c. \end{aligned} \tag{4.10}$$

The specific heat $C = T(\partial S/\partial T)$, and so

$$\begin{aligned} C &= C_0(T), & T &> T_c, \\ C &= C_0(T) + \frac{a^2 T_c}{2B}, & T &< T_c. \end{aligned} \tag{4.11}$$

In Eqs. (4.10) and (4.11), $S_0(T)$ and $C_0(T)$ arise from the regular part $G_0(P, T)$ of the free energy. The resulting jump in the specific heat leads to a λ shape of the curve of $C(T)$, and so this transition is also known as a λ-transition.

Finally, let us consider a magnetic system, for which we identify η^2 with the square of the magnetization M^2, and examine the effect of an external magnetic field H, so that

$$G(T, H, M) = G_0 + AM^2 + BM^4 - MH. \tag{4.12}$$

Since we require that $\partial G/\partial M = 0$, while at $T = T_c$ we found that $A = 0$, it follows that at the critical temperature $M^3 = H/(4B)$, i.e.,

$$M \sim H^{\frac{1}{3}}, \qquad T = T_c. \tag{4.13}$$

The magnetic susceptibility is $\chi = \partial M/\partial H = 1/(2A + 12BM^2)$, so that on substituting the above values for M^2 we find that the magnetic susceptibility χ is given by

$$\begin{aligned} 1/\chi &= 2a(T_c - T), & T &< T_c, \\ 1/\chi &= 4a(T - T_c), & T &> T_c. \end{aligned} \tag{4.14}$$

It follows that the magnetic susceptibility becomes infinite at $T = T_c$.

In all the above considerations we assumed that in Eq. (4.5) the coefficient $\beta(P, T)$ vanishes because of symmetry requirements. If this is not the case, all the above calculations are correct provided that in addition $\beta(P_c, T_c) = 0$ at the critical point, as well as $A(P_c, T_c) = 0$. However, these two conditions mean that the critical point is an isolated point in the T–P plane — a case which is of no special interest.

We note that all the results obtained in Eqs. (4.7)–(4.14) for the temperature dependence of the thermodynamic parameters do not involve the spatial dimensions. This is not correct, as we will see later on. Such a failing should not surprise us, since the mean field approach is based, in particular, on the assumption of analyticity, which is at least questionable for a function which may have singularities.

4.2 First Order Phase Transitions in Landau Theory

So far, we have considered only continuous (second order) phase transitions. However, the Landau mean field theory is also able to describe first order phase transitions, where there is a jump in the order parameter. There are two different ways of doing this:

1. By adding a cubic term to the Landau expansion (4.8), one obtains

$$G = G_0 + a(T - T_c)\eta^2 + C\eta^3 + B\eta^4 . \tag{4.15}$$

The assumption that G as function of η has a minimum, $\partial G/\partial \eta = 0$, leads to

$$\eta_1 = 0, \qquad \eta_2 = -\frac{3C}{8B} \pm \sqrt{\left(\frac{3C}{8B}\right)^2 - \frac{a(T - T_c)}{2B}} . \tag{4.16}$$

A temperature T_0 exists such that

$$G(\eta_1, T_0) = G(\eta_2, T_0) . \tag{4.17}$$

When the system approaches the temperature T_0 from above, the order parameter jumps discontinuously from η_1 to η_2, which means that a first order phase transition occurs there.

2. A first order phase transition can be obtained for negative values of the coefficient B in the Landau expansion (4.8). Then, for a minimum of $G(M)$ to exist, one has to consider the sixth order term in the expansion of $G(M)$,

$$G = G_0 + a(T - T_c)M^2 - |B|M^4 + DM^6 . \qquad (4.18)$$

The conditions for a minimum of Eq. (4.18) lead to

$$M_1 = 0, \qquad M_2 = \frac{1}{\sqrt{3D}}\{|B| + \sqrt{B^2 - 3aD(T - T_c)}\}^{\frac{1}{2}} . \qquad (4.19)$$

By using a procedure identical to that used for the analysis of Eq. (4.17), we conclude that a first order phase transition occurs in this case as well.

4.3 Landau Theory Supplemented with Fluctuations

As we showed earlier in this chapter, the difference between the results of mean field theory and the exact results are due to fluctuations. We now generalize the Landau theory by taking into account these fluctuations. In order to do this, the first step is to use the quasi-hydrodynamical approach, and replace the discrete lattice by a continuum, so that the free energy G and the order parameter M in the case of ferromagnetism become functions of the continuous coordinate r, $G = G(r)$ and $M = M(r)$. Then the global expansion (4.8) has to be replaced by a local one for $G(r)$, while the global free energy and magnetic moment are obtained by integration over the whole d-dimensional system.

For a non-homogeneous system the expansions of the free energy will contain not only the thermodynamic variables but also their derivatives. Assuming that the inhomogeneities are small, one can retain in the expansions only the simple gradient term $(\nabla M)^2$, and neglect higher derivatives. There is no need to multiply this term by a constant since the length scale can be chosen to make this constant unity. Thus, the simplest possible form for the free energy density $G(r)$ is

$$G(r) = G_0 + AM^2 + BM^4 + (\nabla M)^2 . \qquad (4.20)$$

According to Boltzmann's formula, the probability for a fluctuation $P(M)$ is proportional to $\exp[-(G - G_0)/kT]$, and so

$$P(M) \sim \exp\left[-\int d\tau \frac{AM^2 + (\nabla M)^2}{kT}\right] . \qquad (4.21)$$

Here, we ignore the term BM^4 since this was only introduced to prevent $G - G_0$ vanishing at the critical point, where $A = 0$, while the extra term $(\nabla M)^2$ ensures that this does not happen. Thus, we use a Gaussian approximation, i.e., one that involves only quadratic terms in M and ∇M.

In order to calculate the above integral, it is convenient to use the Fourier transform of $M(r)$,

$$M(\mathbf{r}) = \int M_{\mathbf{K}} \exp(i\mathbf{K} \cdot \mathbf{r}) \, d^d K . \qquad (4.22)$$

The radial distribution function $g(r)$ of $M(\mathbf{r})$ is determined by the coefficients $M_{\mathbf{K}}$ [1],

$$g(r) = \frac{\int M(\mathbf{R})M(\mathbf{R} + \mathbf{r}) \, d\mathbf{R}}{\int M(\mathbf{R})M(\mathbf{R}) \, d\mathbf{R}} = \int \langle M_{\mathbf{K}}^2 \rangle \exp(i\mathbf{K} \cdot \mathbf{r}) \, d^d K . \qquad (4.23)$$

According to the Gaussian approximation (4.21), the probability of a fluctuation described by a given set $\{M_{\mathbf{K}}\}$ of Fourier coefficients is

$$P\{M_{\mathbf{K}}\} \sim \exp\left[-\frac{(A + K^2)M_{\mathbf{K}}^2}{kT}\right] \equiv \exp\left(-\frac{M_{\mathbf{K}}^2}{2\langle M_{\mathbf{K}}^2 \rangle}\right) , \qquad (4.24)$$

where $\langle M_{\mathbf{K}}^2 \rangle = kT/[2(A + K^2)]$. On substituting this expression in Eq. (4.23), we find that

$$g(r) = \int \frac{kT \exp(i\mathbf{K} \cdot \mathbf{r})}{2(A + K^2)} d^d K . \qquad (4.25)$$

It follows, e.g. from integration in the complex plane and the residue theorem, that in three dimensions

$$g(r) \sim \frac{\exp(-r\sqrt{A})}{r}, \qquad (4.26)$$

so that the correlation length

$$\xi = \frac{1}{\sqrt{A}} \sim (T - T_c)^{-1/2}. \qquad (4.27)$$

At the critical point $A = 0$, and so at $T = T_c$ the correlation function becomes

$$g(r) \sim \frac{1}{r}. \qquad (4.28)$$

4.4   Correlation Radius and Penetration Length

Let us start from the one-dimensional Schrodinger equation for the order parameter η in a stationary state, which follows from the minimization of the Landau–Ginzburg free energy,

$$-\frac{\hbar^2}{2m}\frac{d^2\eta}{dx^2} + \alpha(T - T_c)\eta + b\eta^3 = 0. \qquad (4.29)$$

For the homogeneous state, $\eta = \eta_0$ is given by

$$\eta_0^2 = \frac{\alpha(T_c - T)}{b}. \qquad (4.30)$$

In order to find the characteristic length in Eq. (4.29), it is convenient to rewrite this equation for the function ψ defined by $\eta = \eta_0\psi$ dependent on the parameter $\xi(T) = \left[\hbar^2/2m\alpha(T_c - T)\right]^{1/2}$. Eq. (4.29) then takes the form

$$-\xi^2(T)\frac{d^2\psi}{dx^2} - \psi + \psi^3 = 0. \qquad (4.31)$$

Therefore, the only length which defines the wave function ψ is the so-called correlation radius $\xi(T)$ for a given temperature T [27].

In order to get the second correlation length, we use the electrodynamic approach, starting from the equation of motion for the coordinate **r** (without electrical resistance) and the expression for the superconducting current **j**

$$m \frac{d^2 \mathbf{r}}{dt^2} = e\mathbf{E}; \quad \mathbf{j} = en \frac{d\mathbf{r}}{dt}; \quad n = \eta_0^2 = \frac{\alpha (T_c - T)}{b}, \qquad (4.32)$$

which results in $d\mathbf{j}/dt = (e^2 n/m) \mathbf{E}$. Inserting the latter equation into the system of Maxwell equations yields

$$\text{curl } \mathbf{H} = \frac{4\pi}{c} \mathbf{j}; \quad \text{curl } E = -\frac{1}{c} \frac{\partial \mathbf{H}}{\partial t}; \quad \text{div } H = \mathbf{0}. \qquad (4.33)$$

In the first equation we have neglected the $\partial \mathbf{E}/\partial t$ term, restricting our consideration to slow processes. Combining Eqs. (4.32) and ((4.33), one easily obtains the dynamic equation for the magnetic field **H**

$$\lambda^2 (T) \nabla^2 \frac{\partial \mathbf{H}}{dt} = \frac{\partial}{\partial t} \frac{\partial \mathbf{H}}{\partial t}, \qquad (4.34)$$

where $\lambda^2 (T) = bmc^2/8\pi e^2 \alpha (T_c - T)$. Equation (4.34) has been obtained from the electrodynamics of a classical charged particle with zero electrical resistance. It turns out, however, that for describing the magnetic field in a superconductor, one has to replace $\partial \mathbf{H}/dt$ in Eq. (4.34) by **H** [27].

Two characteristic lengths, $\xi (T)$ and $\lambda (T)$, determine the correlation radius of the order parameter in the absence of a magnetic field and the penetration length of the magnetic field, respectively. It turns out that the ratio of these two characteristic lenghs (the Ginzburg–Landau parameter) $\kappa = \lambda/\xi$ separates two regions, where the surface tension is positive for $\kappa < 1/2$ and negative for $\kappa > 1/2$. In the latter case (superconductors of the second type), in addition to the characteristic magnetic field $\mathbf{H}_C$, which destroys the superconductivity, there are two other magnetic fields, $\mathbf{H}_1$ and $\mathbf{H}_2$, which are smaller than $\mathbf{H}_C$, such that for the external magnetic field **H** with $\mathbf{H}_1 < \mathbf{H} < \mathbf{H}_2$, a superconductor is in the mixed state containing both superconductor and normal states. The latter phenomenon arises from the negative surface tension, which leads to the appearance of the second phase both for $T < T_C$ and for $T > T_C$.

Below the low critical field $\mathbf{H}_1$, the high-temperature superconductor is still a perfect diamagnet, but in fields just above $\mathbf{H}_1$ magnetic flux does penetrate the material. It is concentrated in well separated "vortices" of size λ, the magnetic penetration depth, carrying
one unit of flux $\Phi_0 \equiv hc/2e$.

The vortices strongly interact with each other, forming highly
correlated stable configurations like the vortex lattice, they can
vibrate and move. The vortex systems in such materials became
an object of experimental and theoretical study early on. Discovery of high T_c materials focused attention to certain particular sit-
uations and novel phenomena within vortex matter physics. They
are "strongly" high-temperature superconductors with $\kappa \sim 100 \gg$
$\kappa_c = 1/\sqrt{2}$ and are "strongly fluctuating" due to high T_c and
large anisotropy in a sense that thermal fluctuations of the vortex
degrees of freedom are not negligible, as was the case in "old" superconductors. In high-temperature superconductors the lower criti-
cal field $\mathbf{H}_1$ and the higher critical field $\mathbf{H}_2$ at which the material
becomes "normal" are well separated $\mathbf{H}_2/\mathbf{H}_1 \sim \kappa^2$ leading to a typical situation $\mathbf{H}_1 \ll \mathbf{H} < \mathbf{H}_2$ in which magnetic fields associated
with vortices overlap, the superposition becoming nearly homoge-
neous, while the order parameter characterizing superconductivity
is still highly inhomogeneous. The vortex degrees of freedom dominate in many cases the thermodynamic and transport properties of
the superconductors.

Thermal fluctuations significantly modify the properties of the
vortex lattices and might even lead to its melting. A new state, the
vortex liquid is formed. It has distinct physical properties from both
the lattice and the "normal" metal. In addition to interactions and
thermal fluctuations, disorder (pinning) is always present, which may
also distort the solid into a viscous, glassy state, so the physical
situation becomes quite complicated leading to rich phase diagram
and dynamics in multiple time scales. A theoretical description of
such systems is a subject of the review article [28].

In the field range $\mathbf{H} \ll \mathbf{H}_2$ vortex cores are well separated and the
magnetic field is inhomogeneous. This model was developed for low T_c
superconductors and subsequently elaborated to describe the high T_c

materials as well. When the distance between vortices is smaller than λ (at fields of several H_2) the magnetic field becomes homogeneous due to overlaps between vortices. In an extreme case of $H \sim H_2$ only small "islands" between core centers remain superconducting, yet superconductivity dominates electromagnetic properties of the material.

4.5 Critical Indices

In the theory of phase transitions, it is very important to describe the behavior of various properties of the system in the vicinity of the critical point. For a differentiable function with a singularity at the critical point, such as $a \sim |T - T_c|^x$, this behavior is characterized by the index x, which is called a critical index. If $x \neq 0$, this critical index is given by $x = \ln(a)/\ln|T - T_C|$, while if $x = 0$ there are two possibilities. In one case, a becomes a constant at the critical point, with the possibility of different values of this constant on the two sides of the transition, which is a jump singularity. In the other case, a exhibits a logarithmic singularity, $a \sim \ln|T - T_c|$. For other properties of the system near and at the critical point, the power exponents of these dependencies are called critical indices. In Table 5.1 in the next chapter, we define a set of critical indices and present their values for the mean field model calculated above, as well as the values of the critical indices for some other models which we analyze there.

4.6 Ginzburg Criterion

Since the difference between the results of mean field theory and the exact ones is due to the fluctuations in the order parameter, we expect that the mean field approximation will be accurate when these fluctuations are sufficiently small. Ginzburg proposed [29] that the mean field theory is applicable when the fluctuations are small in comparison with the thermodynamical values. On using Eq. (4.24) and the fact that $M^2 \sim A$, one finds after integrating over angles that this criterion is fulfilled if

$$\int_0^{K_{\max}} kT \frac{K^{d-1}}{A+K^2} dK < |A|. \tag{4.35}$$

If the dependence of the integral on its upper limit can be neglected, it follows from the substitution $y = K\sqrt{A}$ that this condition becomes

$$(A^{d/2}/A) \int f(y) dy \ll A \tag{4.36}$$

and so $A^{d/2} \ll A^2$. Hence, since A is small in the region of the critical point, the mean field approximation is valid if $d > 4$, so that the upper critical dimension is four. For $d = 4$ there are only logarithmic corrections to the mean field results.

While the above estimates are simple but crude, they still provide us with useful information. If one takes into account the coefficients in the inequality (4.35), and substitutes their values for different types of phase transitions, qualitative estimates can be obtained of the region around the critical point where the mean field theory is valid. It turns out that for the gas–liquid phase transition the mean field theory can be applied up to within ten percent of the critical temperature, while for the superconducting transitions the mean field theory gives the correct results everywhere in the experimentally accessible vicinity of the critical point. The latter result is connected with the large value of the correlation length in low temperature superconductors.

4.7   Wilson's ϵ-Expansion

So far, we have considered the local value $\eta(\mathbf{r})$ associated with the points $\mathbf{r}$ of a lattice, which are a distance a apart and can be regarded as a continuum, so that in our previous analysis we often used integrals rather than sums. Such a coarse-graining procedure does not lead to any problems in hydrodynamics, where a is the only characteristic length, as soon as the condition $r > a$ is satisfied. However, as Wilson pointed out [30], this coarse-graining procedure becomes problematic in the theory of phase transitions, since near the critical point the correlation length ξ appears as an additional characteristic length. Hence, the simple coarse-graining procedure for $r > a$ becomes unjustified unless in addition $r > \xi$, and one has to find some way to treat the range $a < r < \xi$. This region is very important near the critical point, since ξ tends to infinity as the critical point is approached.

In order to deal with this problem, Wilson suggested that, in contrast to the situation for hydrodynamics, the coefficients A and B in Eq. (4.20) depend on the size of the coarse-graining region, which we denote by L. In order to find these dependencies let us consider how the free energy changes when one goes from a region of size L, with $a < L < \xi$, to a region of size $L + \Delta L$. Such a comparison of the free energies in two regions is, in fact, the basis of the renormalization group approach which we will consider in detail in Chapter 6.

Let us consider a value of ΔL so small that it introduces just one additional mode into a system. If we label the modes by their Fourier components K, this means [1] that $V 4\pi K^{d-1} \Delta K / (2\pi)^3 = 1$, or since $K \sim 1/L$ that, for $V = 1$, $\Delta L \sim L^{d+1}$. Let us now write the free energy $G_{L+\Delta L}(P, T)$ in the Landau form (4.20),

$$G_{L+\Delta L}(M) = \int d\tau [G_0 + A_{L+\Delta L} M^2 + B_{L+\Delta L} M^4 + (\nabla M)^2] \quad (4.37)$$

and compare it with that obtained from $G_L(P, T)$ supplemented by an additional fluctuating mode mM_1 with a scaling factor m:

$$G_L(M + mM_1) = \int d\tau \{ G_0 + A_L(M + mM_1)^2$$
$$+ B_L(M + mM_1)^4 + [\nabla(M + mM_1)]^2 \} . \quad (4.38)$$

The contribution of the additional mode mM_1 will be taken into account not by the "hydrodynamic" average but by the correct "Boltzmann" average. Thus, one finds that

$$\exp\left(-\frac{G_{L+\Delta L}}{kT}\right) = \int_{-\infty}^{\infty} dm \exp\left(-\frac{G_L(M + mM_1)}{kT}\right) . \quad (4.39)$$

We restrict our attention to the Gaussian distribution, and so leave in the exponent of the right hand side of Eq. (4.39) only terms quadratic in m, and require that for the fluctuating mode

$$\int M_1(r)dr = 0, \qquad \int M_1^2(r)dr = 1. \tag{4.40}$$

For small ΔL one can write

$$A_{L+\Delta L} = A_L + \frac{dA}{dL}\Delta L, \qquad B_{L+\Delta L} = B_L + \frac{dB}{dL}\Delta L. \tag{4.41}$$

The substitution of Eqs. (4.40) and (4.41) into Eq. (4.39) leads to the equation

$$\frac{dA_L}{dL}\Delta L M^2 + \frac{dB_L}{dL}\Delta L M^4$$
$$= \text{Const} + \ln(1 + A_L L^2 + 6B_L L^2 M^2). \tag{4.42}$$

On expanding the logarithm in (4.42) in a power series and equating the coefficients of M^2 and M^4, we find that

$$\frac{dA}{dL}\Delta L = 3BL^2 - 3ABL^4, $$
$$\frac{dB}{dL}\Delta L = -9B^2 L^4. \tag{4.43}$$

Since $\Delta L \sim L^{d+1}$, this gives in terms of the parameter $\epsilon \equiv 4 - d$,

$$A \sim L^{-\frac{\epsilon}{3}}, \qquad B \sim L^{-\epsilon}. \tag{4.44}$$

Equations (4.44) must be valid for $a < L < \xi$, and so we substitute in them $L = \xi$. On using Eq. (4.27) for ξ and $A = a(T - T_c)$, one finds from Eq. (4.44) that $M \sim [(T - T_c)/T_c]^\beta$ and $\xi \sim (|T - T_c|/T_c)^{-\nu}$ where

$$\beta = \frac{1}{2} - \frac{\epsilon}{6\left(1 - \frac{\epsilon}{6}\right)}; \qquad \nu = \frac{1}{2\left(1 - \frac{\epsilon}{6}\right)}. \tag{4.45}$$

This equation gives the values of the critical indices, which were defined in the last section and are listed in Table 5.1. In this equation, the critical indices are expanded as power series in the small parameter ϵ, which is why the method is known as Wilson's ϵ-expansion. Not by chance is Wilson's article [31] called "Critical phenomena in

3.99 dimensions". If one is brave enough to set the "small" parameter ϵ equal to unity, $\epsilon = 1$, then the one finds that for three-dimensional systems the critical indices are $\beta = 0.3$ and $\nu = 0.6$.

The above approximate calculation, which gives the ϵ-corrections to the mean field critical indices, is a good introduction to the renormalization group approach, where this transformation leading from G_L to $G_{L+\Delta L}$ is applied repeatedly.

4.8 Conclusion

The simplest phenomenological mean field description of phase transitions is obtained by introducing an order parameter η and assuming that close to the critical point the free energy G can be expanded in this order parameter. The requirement that $G(\eta)$ be a minimum enables us to find the critical indices which define the behavior of the thermodynamical parameters near the critical point. As it follows from comparison of Weiss' mean free theory with the Ising model, the main approximation of the mean field theory is the neglect of fluctuations. According to the Ginzburg criterion, this is justified for space dimensions larger than four (the upper critical dimension). The inclusion of fluctuations in the mean field theory is achieved by adding to the free energy the simplest gradient terms, and this enables us to find the temperature dependence of the correlation length and the form of the correlation function at the critical point.

Another type of correction to the mean field theory is associated with the increase of the correlation length on approaching the critical point. These corrections are taken into account by introducing the spatial dependence of the coefficients in an expansion $G = G(\eta)$. This leads to the dependence of the critical indices on the parameter $\epsilon = 4 - d$, which defines the distance from the upper critical dimension $d = 4$. This ϵ-expansion is the simplest form of a renormalization group procedure.

Chapter 5

Scaling

Scaling is a very general and well-known method for describing the response of a system to a disturbance. For physical systems, it is most readily studied in terms of dimensional analysis and the construction of dimensionless parameters, and so we consider this topic first, and start with a few simple examples.

Let us consider first the free fall (i.e., with no air resistance) of a body of mass m under the gravitational force mg. If the body starts from rest, then we expect that its velocity v after falling from a height h will be a function of m, h, and g, $v = f(m, h, g)$. We assume that v has a power-law dependence on these parameters, and so write $v = am^\alpha h^\beta g^\gamma$, where a is a dimensionless constant. In any equation, the dimensions of the quantities on the two sides must be equal. Since $[v] = LT^{-1}, [h] = L, [m] = M$ and $[mg] = MLT^{-2}$, it follows that

$$LT^{-1} = M^\alpha L^\beta (LT^{-2})^\gamma, \qquad (5.1)$$

which has the unique solution $\alpha = 0$, $\beta = \gamma = 1/2$, so that $v = a\sqrt{gh}$. Thus, we have found the functional dependence of v on g and h without having to solve the problem, and can conclude that if the height from which the body falls is multiplied by four then its velocity will be doubled.

Another example, which is very relevant for engineering applications, is the force F acting on a sphere of radius R moving through a fluid of viscosity η with velocity v. We assume that $F = aR^\alpha v^\beta \eta^\gamma$, and once again equate the dimensions of the two sides of this equation. Since $[F] = MLT^{-2}$, $[R] = L$, $[v] = LT^{-1}$, and $[\eta] = ML^{-1}T^{-1}$, it follows that $\alpha = \beta = \gamma = 1$, so that $F = aRv\eta$, which is just Stokes' law. This law permits the prediction of such properties as the air resistance to the motion of cars or aeroplanes from those of model systems with smaller objects and/or lower velocities, although allowance has to be made for the fact that cars and aeroplanes are not spherical. An important point about this example, in contrast to the first one, is that we had to decide (or guess) which of the four parameters in the system, namely the radius, velocity and mass of the object and the viscosity of the medium, was irrelevant, since for mechanical (as opposed to electromagnetic) quantities there are only three independent dimensions, namely mass M, length L and time T.

Scaling is also of considerable importance for presenting theoretical predictions and experimental data for different values of the parameters on a single curve. The basis for such a procedure is that a homogeneous function $f(x, y)$ of x and y of order p is defined by the requirement that

$$f(\lambda x, \lambda y) = \lambda^p f(x, y). \tag{5.2}$$

Then, on choosing $\lambda = 1/x$, one finds that $f(1, y/x) = (1/x)^p f(x, y)$, so that

$$f(x, y) = x^p f\left(1, \frac{y}{x}\right) = x^p \psi\left(\frac{y}{x}\right). \tag{5.3}$$

Equation (5.3) means that a homogeneous function of two variables can be scaled to a function of a single variable, so that by a suitable choice of variables ($f(x, y)/x^p$ and y/x) all of the data collapse onto a single curve.

One can also consider [32] a generalized homogeneous function

$$f(\lambda^a x, \lambda^b y) = \lambda^p f(x, y), \tag{5.4}$$

which for $a = b$ reduces to a standard homogeneous function (5.2).

Table 5.1 Critical indices for the specific heat (α), order parameter (β), susceptibility (γ), magnetic field (δ), correlation length (ν) and correlation function (η).
The indices α, γ, ν and α', γ', ν' refer to $T > T_c$ and $T < T_c$, respectively. The
results for the mean field and 2D Ising models are exact, while those for the 3D
model are approximate.

Exponent	Definition	Condition	Mean Field	2D Ising	3D Ising
α	$C \sim \left\lvert \frac{T-T_c}{T_c} \right\rvert^{-\alpha(-\alpha')}$	$H = 0$	0 (jump)	0 (ln)	~ 0.11
β	$M \sim \left(\frac{T_c-T}{T_c} \right)^{\beta}$	$H = 0;\ T < T_c$	0.5	0.125	~ 0.32
γ	$\chi \sim \left\lvert \frac{T-T_c}{T_c} \right\rvert^{-\gamma(-\gamma')}$	$H = 0$	1	1.75	~ 1.24
δ	$H \sim \lvert M \rvert^{\delta}$	$T = T_c$	3	15	~ 4.82
ν	$\xi \sim \left\lvert \frac{T-T_c}{T_c} \right\rvert^{-\nu(-\nu')}$	$H = 0$	0.5	1	~ 0.63
η	$g(r) \sim \frac{1}{r^{d-2+\eta}}$	$T = T_c$	0	0.25	~ 0.03

5.1 Relations Between Thermodynamic Critical Indices

Critical indices, which were introduced in previous chapters, define
the behavior of the thermodynamic parameters near the critical
point. A list of such indices, with their definitions, is given in
Table 5.1. While the definitions listed there are for a magnetic system, they can easily be formulated for other types of system. For
instance, for the liquid–gas system, the magnetic field H is replaced
by the pressure P, and the magnetic moment M by the specific volume. The different thermodynamic parameters are not independent,
since there are some thermodynamic relations between them, and
these lead to connections between different critical indices.

We now consider two examples of such relations. The well-known
formula [1] linking specific heats in magnetic systems, c_H and c_M (or
c_p and c_v for non-magnetic systems) has the following form:

$$c_H - c_M = \frac{T(\partial M/\partial T)_H^2}{(\partial M/\partial H)_T}. \tag{5.5}$$

In order for a system to be stable, c_M must be positive, and so it
follows from Eq. (5.5) that

$$c_H > \frac{T(\partial M/\partial T)_H^2}{(\partial M/\partial H)_T}, \tag{5.6}$$

or, on substituting the critical indices from Table 5.1,

$$\tau^{-\alpha} > \tau^{2(\beta-1)+\gamma}, \tag{5.7}$$

where $\tau = |T - T_c|/T_c$. Close to the critical point $\tau \ll 1$, and so Eq. (5.7) leads [33] to the following relation between the critical indices

$$\alpha + \beta + 2\gamma \geq 2. \tag{5.8}$$

Another inequality can be obtained as follows [34]. For $T_1 < T_c$,

$$G(T_c, M_1) = G(T_1, M_1) + \int_{T_1}^{T_c} \left(\frac{\partial G}{\partial T}\right)_M dT = G(T_1, M_1) - \int_{T_1}^{T_c} S dT. \tag{5.9}$$

However, since the stability condition requires that

$$c_M \sim -\left(\frac{\partial^2 G}{\partial T^2}\right)_M > 0, \tag{5.10}$$

the derivative $(\partial G/\partial T)$ must be a decreasing function of T, and so $S = -(\partial G/\partial T)_M$ must be an increasing function of T. It follows then from Eq. (5.9) that

$$G(T_c, M_1) \leq G(T_1, M_1) - (T_c - T_1)S(T_1, M_1) \tag{5.11}$$

and

$$G(T_c, M_1) \geq G(T_1, M_1) - (T_c - T_1)S(T_c, M_1). \tag{5.12}$$

For $T < T_c$, a magnetic moment $M(T)$ appears spontaneously, so that $G(T, M(T))$ and $S(T, M(T))$ are fully determined by the temperature T, and so Eq. (5.11) can be rewritten in the form

$$G(T_1) \geq G(T_c) + (T_c - T_1)S(T_1). \tag{5.13}$$

On adding Eqs. (5.13) and (5.12), we obtain

$$G(T_c, M_1) - G(T_c) \leq (T_c - T_1)[S(T_c) - S(T_1)]. \tag{5.14}$$

Now

$$G(T_c, M_1) - G(T_c) \sim M_1^{1+\delta} \sim (T_c - T_1)^{\beta(1+\delta)} \tag{5.15}$$

and

$$[S(T_c) - S(T_1)] \sim (T_c - T_1)^{1-\alpha} \, . \tag{5.16}$$

Thus, Eq. (5.14) leads to

$$\beta(1 + \delta) \geq 2 - \alpha \, . \tag{5.17}$$

The inequalities (5.8) and (5.17) follow from thermodynamics, and so are rigorous. In order to derive additional relations between critical indices one have to use some approximations.

5.2 Scaling Relations

There are several different approximate methods for obtaining scaling relations. We shall follow the method of introducing the concept of a new "block" lattice which can be related to the original "site" lattice, as proposed by Kadanoff [35]. In this chapter we apply the method to thermodynamics, while its application to a Hamiltonian will be the starting point for the renormalization group theory considered in the next chapter.

The Kadanoff construction shown in Fig. 5.1 ("scaling hypothesis") consists of the following three stages.

(1) Divide the original Ising site lattice with lattice constant of unit length into blocks of size L less than the correlation length ξ, i.e., $1 < L < \xi$.

(2) Replace the L^d spins inside each block by a single spin $\mu_i = \pm 1$, the sign of which is determined by that of the majority of the sites inside the block. Since the size L of a block is less than the correlation length ξ, we expect that most of the spins within a block will be of the same sign, so that this majority rule is a very reasonable approximation.

(3) Return to the original site lattice by dividing all lengths by L.

The aim of this procedure is to average over the small degrees of freedom, or in other words to make a coarse graining so as to reduce the number of degrees of freedom of the system. The consequences

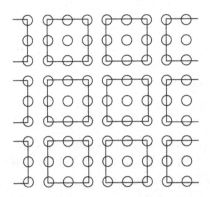

Fig. 5.1   The Kadanoff construction for a square lattice, with blocks of nine sites.

of this idea are not at all trivial. Even if we do not know the exact solution of the problem for either the site lattice or the block lattice, a comparison of these two problems can provide us with valuable information.

Since all the thermodynamic properties of a system depend only on the temperature and the external magnetic field, which determine its free energy, let us consider a site lattice in an external (dimensionless) magnetic field h and in contact with a thermal bath at temperature T, which appears in our analysis in the dimensionless form $\tau = |T - T_c|/T_c$. Let the corresponding parameters for the block lattice be h' and τ'. It is obvious that if there is no external field for the site lattice so that $h = 0$, then there is no such field for the block lattice so that $h' = 0$, and similarly if $\tau = 0$ then $\tau' = 0$. In view of this, and since the single parameter that characterizes the size of the block lattice is L, Kadanoff proposed that $\tau' = L^x \tau$ and $h' = L^y h$.

Let $f(\tau, h)$ be the free energy per site for the site lattice, and $f(\tau', h')$ the free energy per block of the block lattice. Since the free energy of a block is the sum of the free energy of the sites composing it, it follows that in d dimensions

$$L^d f(\tau, h) = f(\tau', h') = f(L^x \tau, L^y h). \qquad (5.18)$$

As can be seen by substitution, a solution of this functional equation is

$$f(\tau, h) = \tau^{d/x} \psi(\tau/h^{x/y}), \qquad (5.19)$$

where ψ is an arbitrary function. This property enables us to find the connection between the critical indices associated with different properties of the system without knowing the function ψ.

The magnetic moment

$$M \sim [\partial f/\partial h]_{h=0} \sim \tau^{d/x}[\tau/h^{1+x/y}]\psi'(z), \qquad (5.20)$$

where $z = \tau/h^{x/y}$. The only way that this can be independent of h as $h \to 0$ is that for $z \longrightarrow \infty$, $\psi'(z) \sim z^{-1-y/x}$, in which case $M \sim \tau^{(d-y)/x}$. Hence, the critical index β, defined in Table 5.1, is given by

$$\beta = \frac{d-y}{x}. \qquad (5.21)$$

It then follows that the magnetic susceptibility as $h \to 0$, i.e., as $z \to \infty$,

$$\chi \sim \partial M/\partial h \sim \tau^{d/x}[\tau/h^{1+x/y}]^2\psi''(z) \qquad (5.22)$$

and thus $\psi''(z) \sim z^{-2-2y/x}$, so that $\chi \sim \tau^{d/x+2-2-2y/x}$, and hence

$$\gamma = \frac{2y-d}{x}. \qquad (5.23)$$

The specific heat as $h \to 0$, i.e., as $z \to \infty$, $C \sim \partial^2 f/\partial\tau^2 \sim \tau^{d/x-2}$, and so

$$\alpha = 2 - \frac{d}{x}. \qquad (5.24)$$

At the critical point, $\tau = 0$, the temperature-independent magnetic moment $M \sim \tau^{d/x}[\tau/h^{1+x/y}]\psi'(z)$ requires $\psi'(z) \sim z^{-(1+d/x)}$ as $z \to 0$. A new critical index δ is defined by $M \sim h^{1/\delta}$ at $\tau = 0$. It then follows that

$$\delta = \frac{y}{d-y}. \qquad (5.25)$$

In order to obtain more critical indices, let us now consider the correlation function $g(r, \tau)$ in the site and block lattices. The first step is to derive the relationship between the average spins $\langle S \rangle$ of the site lattice and $\langle \mu \rangle$ of the block lattice. To do this, we compare the average energy per block of L^d spins in an external magnetic field

h in the site lattice with the energy per block in the corresponding external magnetic field $h' = L^y h$ in the block lattice,

$$hL^d \langle S \rangle = hL^y \langle \mu \rangle \,. \tag{5.26}$$

It follows that $\langle S \rangle \sim L^{y-d} \langle \mu \rangle$, or for the correlation functions g_s and g_μ in the site and block lattices, respectively,

$$g_s(r, \tau) = L^{2(y-d)} g_\mu \left(\frac{r}{L}, L^x \tau \right) \,. \tag{5.27}$$

It can easily be proved by substitution that the solution of the functional equation (5.27) has the following form

$$g_s(r, \tau) = r^{2(y-d)} \Psi(r\tau^{1/x}) \,, \tag{5.28}$$

where Ψ is an arbitrary function. At the critical point, $\tau = 0$, the limiting form of $\Psi(z)$ has to be $\Psi(z) \sim$ constant and so $g_s(r, \tau) \sim r^{2(y-d)}$. On comparing this with the definition of the critical indices in Table 5.1, one finds that

$$2y - d = 2 - \eta \,. \tag{5.29}$$

The characteristic length enters Eq. (5.28) in the form $r\tau^{1/x}$, i.e., for the correlation length ξ one obtains $\xi \sim \tau^{-1/x}$, or

$$\frac{1}{x} = \nu \,. \tag{5.30}$$

Thus, the Eqs. (5.21)–(5.25) and (5.29)–(5.30) define six relations between the critical indices. On eliminating from these relations the unknown parameters x and y, one finds the following four relations between different critical indices

$$\alpha + \beta + 2\gamma = 2, \quad \beta\delta = \beta + \gamma, \quad \gamma = \nu(2 - \eta), \quad \alpha = 2 - d \,. \tag{5.31}$$

The above analysis makes it clear that when the critical indices are defined for $T > T_c$ and $T < T_c$, the calculations can be performed in both cases so that the limiting behavior of the thermodynamic functions is symmetric with respect to T_c, i.e.,

$$\alpha = \alpha', \qquad \nu = \nu', \qquad \gamma = \gamma'. \qquad (5.32)$$

Equations (5.31) and (5.32) show that, within the framework of the scaling hypothesis, there are seven relations between the nine thermodynamic critical indices, i.e., only two indices are independent. These two remaining indices will be found in the next chapter by the use of the renormalization group theory.

Since the scaling hypothesis is based on some postulates, it is worthwhile to check it using the values of critical indices listed in Table 5.1. One can immediately see that the critical relations (5.31) are satisfied for the exact Onsager solution of the two-dimensional Ising model, and (apart from the last dimensions-dependent relation in Eqs. (5.31)) for the mean field theory. It is also clear how important it was to obtain very accurate experimental values of the critical indices.

5.3 Dynamic Scaling

In addition to the above scaling parameters describing a system's static properties near the critical point, its dynamic properties also exhibit singularities, with which are associated dynamic critical indices. Since the dynamic properties describe the approach of a system to its equilibrium state, one cannot consider them in terms of the mean field expression for the static order parameter. Instead, we must consider a system with fluctuations in the order parameter, and so use the free energy used in Chapter 4 in Landau's theory supplemented with fluctuations,

$$G(\eta) = G_0 + A\eta^2 + B\eta^4 + (\nabla\eta)^2. \qquad (5.33)$$

Just as we did in Chapter 4, we ignore in what follows the term in η^4, since the $(\nabla\eta)^2$ term is sufficient to make the free energy depend on η even for $T = T_c$, when $A = 0$. For the static equilibrium system, where obviously η is independent of time, we required that $\partial G/\partial\eta = 0$ and $\partial^2 G/\partial\eta^2 > 0$, and so we postulate that when the system is not far from equilibrium,

$$\frac{d\eta}{dt} = -\Gamma \frac{\delta G}{\delta \eta} = -\Gamma \left[\frac{\partial G}{\partial \eta} - \nabla \frac{\partial G}{\partial (\nabla \eta)} \right], \qquad (5.34)$$

namely, the Landau–Khalatnikov equation. A simple justification of this equation lies in the fact that in equilibrium both sides of this equation vanish, and it is natural to assume that they are proportional for states which are close to equilibrium. We can choose the time scale so that $\Gamma = 1/2$, and so obtain the equation

$$\frac{d\eta}{dt} = \nabla^2 \eta - A\eta. \qquad (5.35)$$

As in the previous chapter, we use the Fourier transform η_K of η, so that

$$\eta_K(t) = \eta_K(0) \exp[-t(A + K^2)] = \eta_K(0) \exp\{-t[a(T_c - T) + K^2]\}, \qquad (5.36)$$

where we have used the mean field value of A as derived in the Landau theory. If one denotes by τ_K the relaxation time for mode K so that $\eta_K(t) = \eta_K(0) \exp[-t/\tau_K]$ then we see that for $K = 0$ the relaxation time $\tau_0 \sim 1/(T_c - T)$, which tends to infinity as T approaches the critical temperature T_c, while for other values of K, $\tau_K \sim 1/K^2$ as $T \to T_c$. Thus, the approach to equilibrium is very slow near the critical temperature, a phenomenon known as critical slowing down. Great care is required to ensure that measurements of thermodynamic properties of the system in this temperature region are performed on systems in equilibrium.

The above results are for the mean field approximation, but in complete analogy with Eq. (5.28) we can obtain similar results for the more general case. To do this, we write

$$\tau_K = f(K, \tau) = \tau^z \Psi(K\tau^{-\nu}), \qquad (5.37)$$

where τ is the reduced temperature and z is the dynamic scaling exponent. In order for τ_K to be proportional to a power of τ, $\Psi(0)$ must be a constant. It then follows that for $K = 0$, $\tau_0 \sim \tau^z$. For $K \neq 0$, the argument of Ψ tends to infinity as $\tau \to 0$, i.e., as $T \to T_C$ and so τ_K can only remain finite if $\Psi(y) \sim y^{z/\nu}$, in which case $\tau_K \sim K^{z/\nu}$.

In the mean field case, we found that $\tau_0 \sim 1/\tau$ as $\tau \to 0$ and that $\tau_K \sim 1/K^2$, so that $z = -1$ and $z/\nu = -2$, which leads to the result $\nu = 1/2$, in agreement with what we found earlier. Therefore, in addition to the thermodynamic critical indices, a new critical index z appears in the description of dynamic phenomena.

5.4 Conclusion

The problem of calculating the critical indices which define the behavior of thermodynamic functions near the critical points is very much simplified by the use of the scaling hypothesis. The basic idea is as follows. Close to the critical point the correlation length ξ becomes much larger than the distance a between particles, and there are many "block" lattices of size L containing L^d spins, such that $a < L < \xi$. It is natural to assume that in each block almost all spins would be or up or down, just as in the original site lattice. Therefore, the critical phenomena will be similar in all block lattices, provided that the external parameters, say temperature and magnetic field, are scaled, as compared with the original site lattice, by a simple power-law dependence on L. Although the two indices which define such a change in the external parameters remain unknown, all other critical indices can be easily expressed in terms of these two indices. The scaling relations deduced from this hypothesis are satisfied by the results of the exact solution of the two-dimensional Ising model and by those of the mean field theory. The simplified examination of dynamic phenomena near the critical points requires the knowledge of an additional (dynamic) critical index, and results in the slowing down of the approach to equilibrium.

Chapter 6

The Renormalization Group

The aim of renormalization group theory (RG) is to find the critical indices, including the indices x and y, which were still unknown after using the scaling hypothesis in the previous chapter. RG is a group of transformations (or more rigorously a semi-group, since in general there is no inverse transformation) from a site lattice with lattice constant unity and the Hamiltonian H to a block lattice with lattice constant L and the Hamiltonian H' without changing the form of the partition function,

$$Z(H, N) = Z(H', N/L^d).$$ (6.1)

In fact, this idea was already used in Eq. (5.18), where we compared the free energies of site and block lattices. In this chapter, we consider the connection between Hamiltonians rather than between free energies.

6.1 Fixed Points of a Map

The key idea in the use of the renormalization group is that the transition from H to H' can be regarded as a rule, $K' = f(K)$, for obtaining the parameters K' of the Hamiltonian H' of a block lattice from those parameters K of the Hamiltonian H of the site lattice. This process can be repeated, with the lattice of small blocks being treated as a site lattice for a lattice of larger blocks, so that for the latter $K'' = f(K') = f[f(K)]$, and so on indefinitely for larger and larger blocks until their size reaches the correlation length. This is an example of a map, i.e., a uniquely defined rule relating two successive members of a sequence, x_n and x_{n+1} by means of $x_{n+1} = f(x_n)$. In such a map, it can happen that as $n \to \infty$ this process may stop at some stage, when $x^* = f(x^*)$, in which case x^* is called a fixed point of the map.

For instance, one of the simplest examples of a non-linear map is the logistic map [37]

$$x_{n+1} = f(x_n) = \mu x_n (1 - x_n), \ 0 < x < 1, \quad \mu > 1 \qquad (6.2)$$

so that the fixed points are the roots of the equation $x^* = \mu x^* (1 - x^*)$. Since this is a quadratic map, it has two fixed points, namely $x^* = 0$ and $x^* = 1 - 1/\mu$. This situation is shown in Fig. 6.1 where the fixed points are the intersection of the graph of the parabola, $f(x_n)$, and of the diagonal of the unit box. Starting from some initial point x_0 we make a vertical shift to the point $x_1 = f(x_0)$ lying on the parabola, which (after the horizontal shift to the diagonal) will be the starting point x_1 for the next approximation, and so on, reaching finally the stable fixed point $x^* = 1 - 1/\mu$. In order to examine the stability of the fixed points, we introduce a perturbation ξ_n from x^* in the form

$$x_n = x^* + \xi_n \qquad (6.3)$$

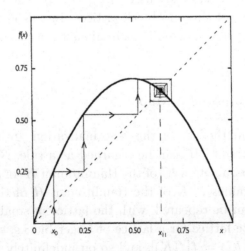

Fig. 6.1   Approach to the fixed point for the logistic map.

and substitute it in Eq. (6.2). On expanding $f(x_n)$ in a Taylor series around x^* and keeping only the terms linear in the small parameters ξ_n, we then find that

$$\xi_{n+1} = \left(\frac{df}{dx}\right)_{x=x^*} \xi_n \, . \tag{6.4}$$

The small distance ξ_n from the fixed point will decrease at each stage if $|df/dx|_{x=x^*} < 1$, and in this case the fixed point will attract the set of points x_n so that the fixed point will be stable. On the other hand, if $|df/dx|_{x=x^*} > 1$, the fixed point is unstable. For the logistic map of Eq. (6.2), $(df/dx) = \mu(1 - 2x)$, and since $\mu > 1$ it is larger than unity for $x^* = 0$ and smaller than unity for $x^* = 1 - 1/\mu$, so that the latter fixed point is the stable one.

6.2 Basic Idea of the Renormalization Group

The above procedure is the basic idea of the renormalization group. Assuming that a fixed point is reached at some step of the iteration procedure for

$$K'' = f(K') = f[f(K)] = \cdots \tag{6.5}$$

namely $K^* = f(K^*)$, where $K = J/(kT)$, one finds the value of $K^* = J/(kT_C)$ which defines the critical temperature, as discussed below. This estimate will be more precise the closer the size of the block lattice to the correlation length, i.e., the more iteration steps that are used.

A very important property of the RG procedure is that it enables one to find not only the critical temperature but also the critical indices. Indeed, on using the RG relation $K' = f(K)$ near a fixed point, one obtains the equation

$$K' - K^* = f(K) - f(K^*) \sim \left(\frac{df}{dK}\right)_{K=K^*} (K - K^*) + \cdots . \tag{6.6}$$

Instead of trying to solve the problem for a single lattice, which is usually impossible, we compare the solution of the problem for lattices of different sizes. Since the free energy per particle $g(K)$

must be unaltered, for a block of L^d sites in d dimensions $L^d g(K_1) = g(K_2)$, or for the partition function $Z(K_1, N) = Z(K_2, N/L^d)$. As noted previously, we can only expect the properties of these lattices to be similar if the size of the blocks is less than the correlation length of the system. However, at a fixed point this similarity applies for blocks of arbitrary size. Therefore, we identify the fixed point with the critical point of a phase transition, since only there is the correlation length infinite. The existence of such a fixed point shows that a phase transition takes place. If a value K' of K is close to the fixed point K^*, we can rewrite Eq. (6.6) for $T = J/kK$ in the form

$$T' - T^* \sim \left(\frac{df}{dK}\right)_{K=K^*} (T - T^*). \qquad (6.7)$$

On comparing Eq. (6.7) with the main scaling relation $\tau' = L^x \tau$, where $\tau = |T - T_c|/T_c$, one finds that

$$x = \ln\left[\left(\frac{df}{dK}\right)_{K=K^*}\right]/\ln(L). \qquad (6.8)$$

If there is more than one external parameter, say K and H, then the function f will depend on more than one parameter, $f = f(K, H)$, and so we will have two RG equations

$$K' = f_1(K, H), \qquad H' = f_2(K, H), \qquad (6.9)$$

and the fixed points are defined by the solutions of equations

$$K^* = f_1(K^*, H^*), \qquad H^* = f_2(K^*, H^*). \qquad (6.10)$$

On repeating all the procedures leading to (6.7), one finds that

$$T' - T^* \sim \left(\frac{df_1}{dK}\right)_{K=K^*} (T - T^*) + \left(\frac{df_1}{dH}\right)_{K=K^*} (H - H^*),$$

$$H' - H^* \sim \left(\frac{df_2}{dK}\right)_{K=K^*} (T - T^*) + \left(\frac{df_2}{dH}\right)_{K=K^*} (H - H^*). \qquad (6.11)$$

Finally, on diagonalizing the 2×2 matrix and writing the diagonal elements in the form L^x and L^y, one finds the critical indices x and y.

It should be noted that the procedure described here is the simplest real space form of the renormalization techniques. In many applications, the renormalization group technique is used in momentum space rather than in real space.

6.3   RG: 1D Ising Model

We start our examination of the application of the RG method by considering the simplest possible system, namely the 1D Ising model, even though we know that it does not exhibit a phase transition. For the original site lattice, the partition function (3.1) for the one-dimensional Ising model (3.4), with $K = J/(kT)$, has the form

$$Z(K, N) = \sum \exp[K(S_1 S_2 + S_2 S_3 + \cdots + S_{N-1} S_N)] \qquad (6.12)$$

where the summation is performed over all neighboring pairs among the N spins. In the block lattice, we omit all the spins associated with even-numbered sites and replace two spins by a single one, so that there are $N/2$ spins (which we referred to previously as μ_j, but in this case we write as S_{2j-1}) for which we have to find a partition function $Z(K', N/2)$ which involves summing only over the odd spins. The first two terms in the exponents of Eq. (6.12), after summation over $S_2 = \pm 1/2$, can be written in the following form

$$\exp\left[\frac{K}{2}(S_1 + S_3)\right] + \exp\left[-\frac{K}{2}(S_1 + S_3)\right] = F \exp[K'(S_1 S_3)], \quad (6.13)$$

and the same procedure can be performed for each two subsequent terms. In this way, we exclude from the sum all the spins with even indices. Thus, we can compare the partition functions for the site and block lattices, and write

$$Z(K, N) = F^{\frac{N}{2}} Z(K', N/2). \qquad (6.14)$$

Equation (6.13) gives for $S_1 = S_3$ and $S_1 = -S_3$

$$\begin{aligned} 2\cosh(K) &= F \exp(K'), \\ 2 &= F \exp(-K'), \end{aligned} \qquad (6.15)$$

so that

$$K' = \frac{1}{2} \ln[\cosh(K)], \qquad F = 2\sqrt{\cosh(K)}. \qquad (6.16)$$

Let us now compare the free energy per site, $f_1(K)$, for the site lattice and that, $f_2(K')$, for the block lattice. Since

$$f_1(K) = (1/N) \ln[Z(K, N)] \quad \text{and} \quad f_2(K') = (2/N) \ln[Z(K', N/2)]$$
$$(6.17)$$

it follows from Eq. (6.14)–(6.16) that

$$f(K') = 2f(K) - \ln\left(2\sqrt{\cosh(K)}\right) . \qquad (6.18)$$

Moreover,

$$K' = \frac{1}{2} \ln\left[\{\exp(K) + \exp(-K)\}/2\right] \qquad (6.19)$$
$$< \frac{1}{2} \ln\left[\exp(K)/2\right] = K - \frac{1}{2} \ln 2 ,$$

i.e., K' is smaller than K, and so by repeatedly doubling the size of the blocks one can proceed to a very small value of interactions between spins far away from each other. Therefore, on going in the reverse direction from small to large interactions, one can start with a very small value of the interaction, say $K' = 0.01$, so that $Z(K') \approx 2^N$ and $f \approx \ln(2)$. Using the above equations, one can find the renormalized interaction and the appropriate free energy. It turns out [36] that it takes eight steps to proceed from $K' = 0.01$ to the exact result.

The one-dimensional Ising model does not show a phase transition at a finite temperature, and our analysis was only designed to show by means of a simple example how the RG method works. We now turn to the two-dimensional Ising model, for which we have seen in Chapter 3 that a phase transition does occur.

6.4   RG: 2D Ising Model for the Square Lattice (1)

For the 2D Ising model on a square lattice, the partition function contains the interaction between a spin and its four nearest neighbors, so that the analysis is more complicated. However, we start by applying the same procedure as for the 1D Ising lattice with $S = \pm\frac{1}{2}$, namely we form the block lattice by omitting the nearest neighbors of

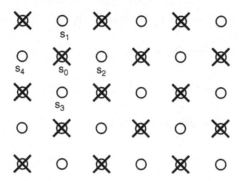

Fig. 6.2 Ising model on a square lattice with alternate sites removed.

half the sites, as shown in Fig. 6.2, and summing over the interactions with the spins of the sites that we omitted. Since the spin S_0 located at the site $(0,0)$ interacts with the spins S_1, S_2, S_3, S_4, located at the sites $(0,1), (1,0), (0,-1), (-1,0)$ respectively, we must consider the terms in the partition function arising from $S_0 = \pm\frac{1}{2}$ with all the 16 possible values of (S_1, S_2, S_3, S_4). As we will see immediately, these 16 configurations are divided into four different groups which will give four different equations instead of the two equations (6.15) for the one-dimensional lattice. In order to satisfy these equations we are forced to introduce three different interactions in the partition function of a block lattice, i.e., to replace the cell of the site lattice shown in Fig. 6.2 by the following interactions in a block lattice

$$\exp[K(S_1 + S_2 + S_3 + S_4)] + \exp[-K(S_1 + S_2 + S_3 + S_4)]$$

$$= F\{\exp[(K_1/2)(S_1 S_2 + S_2 S_3 + S_3 S_4 + S_4 S_1)$$

$$+ K_2(S_1 S_3 + S_2 S_4) + K_3 S_1 S_2 S_3 S_4]\} \tag{6.20}$$

(The reason for using $K_1/2$ rather than K_1 in the above equation is that the products of spins such as $S_1 S_2$ appear again in some other term of the partition function for the block lattice.) In order to determine the values of the four unknowns f, K_1, K_2, K_3 we equate the two sides of this equation for the four distinct values of (S_1, S_2, S_3, S_4), and take the logarithms of each side. For

$(S_1, S_2, S_3, S_4) = (\frac{1}{2}, \frac{1}{2}, \frac{1}{2}, \frac{1}{2})$, we find from Eq. (6.20) that

$$\ln[2\cosh(2K)] = \ln F + K_1/2 + K_2/2 + K_3/2\,. \tag{6.21}$$

Similarly, for $(S_1, S_2, S_3, S_4) = (\frac{1}{2}, \frac{1}{2}, \frac{1}{2}, -\frac{1}{2})$ one obtains

$$\ln[2\cosh(K)] = \ln F - K_3/4\,, \tag{6.22}$$

while $(S_1, S_2, S_3, S_4) = (\frac{1}{2}, \frac{1}{2}, -\frac{1}{2}, -\frac{1}{2})$ leads to

$$\ln 2 = \ln F - K_2/2 + K_3/4\,, \tag{6.23}$$

and $(S_1, S_2, S_3, S_4) = (\frac{1}{2}, -\frac{1}{2}, \frac{1}{2}, -\frac{1}{2})$ to

$$\ln 2 = \ln F - K_1/2 + K_2/2 + K_3/4\,. \tag{6.24}$$

This set of four linear equations for the four unknowns has the unique solution

$$K_2/4 = K_1/8 = \frac{1}{8}\ln[\cosh(2K)]\,,$$

$$K_3/4 = K_2/4 - \frac{1}{2}\ln[\cosh(K)]\,, \tag{6.25}$$

$$F = 2[\cosh(K)]^{\frac{1}{2}}[\cosh(2K)]^{\frac{1}{8}}\,.$$

In contrast to the one-dimensional Ising model, for the two-dimensional Ising model, as well as in some other cases, the resulting interactions in a block lattice are more complicated than those in the original lattice, and so they are not suitable for an exact RG analysis. To proceed, one must use some approximation. The simplest approximation is to neglect K_2 and K_3, but then we come back to the equations considered in the one-dimensional case, where there is no phase transition. To obtain a physically meaningful result, we assume that the four-spin interaction is so weak that it can be ignored. The terms in K_1 and K_2 correspond, respectively, to interactions between nearest neighbors and next-nearest neighbors in the block lattice, and both favor the spins being parallel, and so we write

$K_1 + K_2 = K'$. Thus,

$$K' = (1/2)\ln[\cosh(2K)]. \tag{6.26}$$

Finally, as discussed above, a phase transition occurs when K is a fixed point of the transformation, i.e., $K = K' = K_c$. For the 2D Ising model, this leads to $K_c = 0.50698$, while Onsager's exact result was $K_c = 0.44069$ [15], so that the error is about 13%. This is not too bad for such an approximate theory, and suggests that its basic physical ideas are correct.

6.5   RG: 2D Ising Model for the Square Lattice (2)

Let us now choose a different form of the block lattice, in which we combine five neighboring spins into a block as shown in Fig. 6.3. The spin μ_a of the block lattice is determined by the majority rule,

$$\mu_a = sgn \sum_{i=1}^{5} S_i^a. \tag{6.27}$$

As can be seen, the block lattice is also a square lattice with a distance between blocks of $L = \sqrt{5}$. For $\mu_a = 1$, there are sixteen possible values of S^a, which can be grouped into the six different configurations shown in Fig. 6.4, having characteristic energies zero,

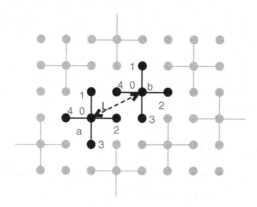

Fig. 6.3   Ising model on a square lattice with five sites in a block.

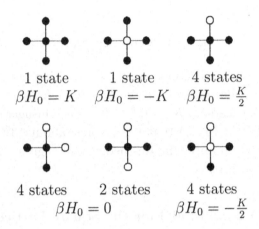

1 state $\quad$ 1 state $\quad$ 4 states

$\beta H_0 = K \quad \beta H_0 = -K \quad \beta H_0 = \frac{K}{2}$

4 states $\quad$ 2 states $\quad$ 4 states

$\beta H_0 = 0 \qquad \beta H_0 = -\frac{K}{2}$

Fig. 6.4 The different spin configurations for the five sites in a block on the square lattice, where black and white circles represent spins of opposite signs. The number of states in each configuration and their energy are shown.

$\pm K/2$, and $\pm K$. For $\mu_a = -1$ one obtains the same energies as for $\mu_a = 1$ with signs reversed.

In order to calculate the partition function Z_μ for the Hamiltonian H_μ of the block lattice it is convenient to write $H_\mu = H_0 + V$, where H_0 contains the interactions between the spins within a block and V those between spins on adjacent blocks, so that

$$Z_\mu = \sum \exp[-\beta H_\mu] = \sum \exp(-\beta H_o) \exp(-\beta V)$$

$$= \sum \exp(-\beta H_o) \left[\frac{\sum \exp(-\beta H_o) \exp(-\beta V)}{\sum \exp(-\beta H_o)} \right], \quad (6.28)$$

where the first factor in the final expression in Eq. (6.28) is related to the independent blocks, and therefore reduces to a multiple of the partition function $Z_0(K)$ of a single block

$$\sum \exp(-\beta H_o) = [Z_0(K)]^N. \qquad (6.29)$$

The sixteen configurations that contribute to $Z_0(K)$, which are shown in Fig. 6.4 for $\mu_a = 1$, lead to

$$Z_0(K) = 2\cosh(K/4) + 8\cosh(K/8) + 6. \qquad (6.30)$$

The second factor in Eq. (6.28) is the weighted average of $\exp(-\beta V)$ with weight function $\exp(-\beta H_o)$, which will be denoted by $\langle \exp(-\beta V) \rangle$.

The formula (6.28) is still exact. However, we only evaluate $\langle \exp(-\beta V) \rangle$ approximately, by the so-called cumulant expansion. In this one writes

$$< e^W >=< 1 + W + \frac{W^2}{2} + \frac{W^3}{\hat{\ }} + ... >= 1+ <W> + < \frac{W^2}{2} > +....$$
(6.31)

For $|x| < 1$, $\ln(1 + x) = x - x^2/2 + \cdots$, and we only retain terms up to order W. Then, $\ln < e^W > \approx < W >$, so that

$$< e^W > \approx \exp < W > .$$
(6.32)

On substituting Eqs. (6.29) and (6.32) in Eq. (6.28), we obtain

$$\beta H_\mu = N \ln[Z_0(K)] - < \beta V > .$$
(6.33)

The function $< \beta V >$ connects adjacent spins of different blocks. As one can see from Fig. 6.3, the interaction between nearest neighbor spins on two adjacent blocks a and b contains only three terms,

$$\beta V_{ab} = K(S_2^a S_4^b + S_1^a S_4^b + S_2^a S_3^b).$$
(6.34)

Since the spins S^a and S^b are on different blocks, while the Hamiltonian H_0 neglects the interactions between blocks, on averaging with respect to H_0 we find that

$$< \beta V_{ab} > = 3K < S_2^a >< S_4^b >$$
(6.35)

In evaluating $< S_2^a >$ the configurations with βH_0 which contain $S_2^a = +\frac{1}{4}$ and $S_2^a = -\frac{1}{4}$ cancel out, and so one finds for $\mu_a = 1$ that

$$\langle S_2^a \rangle = \mu_a \left[\frac{2\cosh(K) + 4\cosh(K/2)}{2\cosh(K) + 8\cosh(K/2) + 6} \right] .$$
(6.36)

An analogous treatment for $\mu_a = -1$ shows that equation (6.36) applies also for this case. From Eqs. (6.35) and (6.36), we find

Table 6.1   The critical value of K, $K_c = kT_c/J$, and the critical index x for different forms of renormalization group theory applied to the Ising model for a square lattice. The column Half is for the blocks of RG-2D(1) with half the spins removed, and the column 5-spin is for the blocks of 5 spins of RG-2D(2), while the column Exact presents the exact results obtained by Onsager [15].

	Half	5-spin	Exact
K_C	0.5070	0.5931	0.4406
x	1.07	0.901	1

that

$$< \beta V_{ab} > = \frac{3}{4} K \left(\frac{\cosh(K) + 2\cosh(K/2)}{\cosh(K) + 4\cosh(K/2) + 3} \right)^2 \mu_a \mu_b . \qquad (6.37)$$

Hence, to first order in V, $\beta H_\mu = \text{Const} + K' \sum_{a,b} \mu_a \mu_b$, where

$$K' = \frac{3}{4} K \left(\frac{\cosh(K) + 2\cosh(K/2)}{\cosh(K) + 4\cosh(K/2) + 3} \right)^2 . \qquad (6.38)$$

In Table 6.1 we show the results of calculations of the critical value of K, $K_c = kT_c/J$, and of the critical index x defined by $(dK'/dK)_{K*} = L^x$, for the square lattice, obtained by the RG method with the two different blocks considered above, and also the exact results of Onsager [15]. By using cumulant expansions, one can also perform the renormalization group calculations for the two-dimensional triangular lattice [38], and the results obtained are close to the exact results of Onsager.

6.6   Conclusion

The same idea of a comparison of site and block lattices which was used in the previous chapter for the formulation of the scaling hypothesis can be used also as a basis for the renormalization group technique. The RG transformations, which were originally used in quantum field theory, are suitable for the phenomenon of phase transitions, where the main feature is the long-range correlations. Therefore, on going from small to larger block lattices, we gradually exclude the small-scale degrees of freedom. For each new block lattice one has to construct effective interactions, and find their connection with the interactions of the previous lattice. These steps are carried out repeatedly. The fixed points of RG transformations define the critical parameters, and series expansions near the critical point provide the values of the critical indices. We note that the fixed points are a property of transformations which are not particularly sensitive to the original Hamiltonian, and this is the basis of the idea of universality which will be considered in Chapter 8.

Chapter 7

Phase Transitions in Quantum Systems

Our treatment of phase transitions so far has considered only phenomenological properties and model systems, without any microscopic analysis of real physical systems. In this chapter, we extend our treatment by considering many-body quantum systems, in which the two best-known phase transitions are superfluidity and superconductivity. Since these phenomena are associated with the symmetry of the wave function and quantum statistical physics, we start with a brief review of these two topics.

7.1 Symmetry of the Wave Function

According to Heisenberg's uncertainty principle, the product of the uncertainty in the position, Δr, and that in its momentum, Δp, can never be smaller than Planck's constant, $\Delta r \Delta p \geq h$. As a result, if we could identify all the particles of the same type, for instance electrons, in a system at some given moment by the exact position of each one, we would have no idea what their momenta are, and so where each one would be at any subsequent time. Hence, we could no longer distinguish between them. As a result of this indistinguishability of identical particles, the amplitude of the wave function describing them must be left unchanged if the labels of the particles are permuted. Thus, for a pair of identical particles with coordinates (position and spin) ξ_1 and ξ_2, the wave function $\psi(\xi_1, \xi_2)$ describing their state can only be multiplied by a phase factor when the particles are interchanged,

$$\psi(\xi_2, \xi_1) = e^{i\alpha}\psi(\xi_1, \xi_2) \,. \tag{7.1}$$

On interchanging the particles once again, one finds that

$$\psi(\xi_1, \xi_2) = e^{i\alpha}\psi(\xi_2, \xi_1) = e^{2i\alpha}\psi(\xi_1, \xi_2) \,, \tag{7.2}$$

and so, since one has now returned to the original system, $e^{2i\alpha}$ must equal unity, so that $e^{i\alpha} = \pm 1$. Hence $\psi(\xi_2, \xi_1) = \pm\psi(\xi_1, \xi_2)$, i.e., the wave function is either symmetric or anti-symmetric with respect to the interchange of particles. It is known from relativistic quantum theory that particles with integer spin, which are called bosons, have symmetric wave functions, and those with half-integer spin, which are called fermions, have anti-symmetric wave functions. In the latter case, two particles cannot have the same coordinates, since the requirement that $\psi(\xi_1, \xi_1) = -\psi(\xi_1, \xi_1)$ can only be satisfied if $\psi(\xi_1, \xi_1) = 0$, and this property is known as the Pauli exclusion principle.

If the wave function of a pair of particles $\psi(\xi_1, \xi_2)$ is expressed in terms of the product of single particle wave functions $\phi_1(\xi)$ and $\phi_2(\xi)$, then the normalized wave function is

$$\psi(\xi_1, \xi_2) = \frac{1}{\sqrt{2}}[\phi_1(\xi_1)\phi_2(\xi_2) \pm \phi_1(\xi_2)\phi_2(\xi_1)] \,, \tag{7.3}$$

where the signs plus and minus correspond to bosons and fermions, respectively. For a system of N bosons, the symmetric wave function $\psi_s(\xi_1, \xi_2, \ldots, \xi_N)$ is given by

$$\psi_s(\xi_1, \xi_2, \ldots, \xi_N) = C_N \sum P\phi_1(\xi_1)\phi_2(\xi_2) \cdots \phi_N(\xi_N) \,, \tag{7.4}$$

where the sum is over all possible permutations P of the coordinates ξ_j and C_N is a normalization constant. For N fermions the anti-symmetric wave function $\psi_a(\xi_1, \xi_2, \ldots, \xi_N)$ can be expressed as a

Slater determinant

$$\psi_a(\xi_1, \xi_2, \ldots, \xi_N) = A_N \begin{Vmatrix} \phi_1(\xi_1) & \cdots & \cdots & \phi_1(\xi_N) \\ \cdots & \cdots & \cdots & \cdots \\ \phi_N(\xi_1) & \cdots & \cdots & \phi_N(\xi_N) \end{Vmatrix}, \quad (7.5)$$

where A_N is a normalization constant. For this wave function, if two particles are in identical states then two rows of the determinant are identical, so that the wave function vanishes, in accordance with the exclusion principle.

7.2 Exchange Interactions of Fermions

Let us now consider a pair of fermions and distinguish between their spatial coordinates $\mathbf{r}_j$ and their spin coordinates σ_j. In the absence of spin–orbit interaction, we can express the anti-symmetric function $\psi_a(\xi_1, \xi_2)$ as a product of a function $\Phi(\mathbf{r}_1, \mathbf{r}_2)$ of the space coordinates and one $K(\sigma_1, \sigma_2)$ of the spin coordinates, i.e.,

$$\psi_a(\xi_1, \xi_2) = \Phi(\mathbf{r}_1, \mathbf{r}_2) K(\sigma_1, \sigma_2). \quad (7.6)$$

The condition that ψ_a be antisymmetric with respect to an interchange of the coordinates ξ_j then requires that one of these functions be symmetric and one anti-symmetric. If one expresses the function $K(\sigma_1, \sigma_2)$ in terms of the product of single particle spin functions, $\chi_1(s_1)$ and $\chi_2(s_2)$, then the symmetric state (triplet), with total spin $S = 1$, can be formed in three different ways,

$$\chi_1(s_1)\chi_1(s_2), \quad \chi_2(s_1)\chi_2(s_2) \quad \text{and} \quad \chi_1(s_1)\chi_2(s_2) + \chi_1(s_2)\chi_2(s_1), \quad (7.7)$$

while the anti-symmetric state (singlet), with total spin $S = 0$, has the form

$$\chi_1(s_1)\chi_2(s_2) - \chi_1(s_2)\chi_2(s_1). \quad (7.8)$$

For the triplet state the spatial wave function, $\Phi_-(\mathbf{r}_1, \mathbf{r}_2)$, is anti-symmetric, whereas for the singlet state, $\Phi_+(\mathbf{r}_1, \mathbf{r}_2)$, it is symmetric, where the wave functions are

$$\Phi_{\pm}(\mathbf{r}_1, \mathbf{r}_2) = \frac{1}{\sqrt{2}}[\phi_1(\mathbf{r}_1)\phi_2(\mathbf{r}_2) \pm \phi_1(\mathbf{r}_2)\phi_2(\mathbf{r}_1)]. \qquad (7.9)$$

According to first order perturbation theory, an interaction potential of the form $V(\mathbf{r}_1 - \mathbf{r}_2)$, such as the Coulomb potential, changes the energy of the states $\Phi_{\pm}$ by

$$\Delta E = \int |\Phi_{\pm}(\mathbf{r}_1, \mathbf{r}_2)|^2 V(\mathbf{r}_1 - \mathbf{r}_2) d\tau_1 d\tau_2. \qquad (7.10)$$

Substituting Eq. (7.9) into Eq. (7.10), one can rewrite Eq. (7.10) in the form

$$\Delta E = A \pm J_1, \qquad (7.11)$$

where A is the Coulomb interaction between the particles and J_1 is the exchange interaction,

$$
\begin{aligned}
A &= \int d\tau_1 d\tau_2 |\phi_1(\mathbf{r}_1)|^2 |\phi_2(\mathbf{r}_2)|^2 V(\mathbf{r}_1 - \mathbf{r}_2), \\
J_1 &= \int d\tau_1 d\tau_2 \phi_1(\mathbf{r}_1)\phi_2(\mathbf{r}_2)\phi_1^*(\mathbf{r}_2)\phi_2^*(\mathbf{r}_1) V(\mathbf{r}_1 - \mathbf{r}_2).
\end{aligned}
\qquad (7.12)
$$

It follows that $\Delta E = A + J_1$ for the singlet state, $S = 0$, while $\Delta E = A - J_1$ for the triplet state, $S = 1$, so that the physically important difference between these two energies is just $2J_1$. This difference can be expressed in terms of the spin operators $\mathbf{s}_1$, $\mathbf{s}_2$ and $\mathbf{S} = \mathbf{s}_1 + \mathbf{s}_2$ or $S^2 = s_1^2 + s_2^2 + 2\mathbf{s}_1 \cdot \mathbf{s}_2$. Since $s_1^2 = s_2^2 = \frac{1}{2}\left(\frac{1}{2} + 1\right) = \frac{3}{4}$ and $S^2 = S(S+1)$ equals zero for the singlet state and two for the triplet states, the eigenvalues of the operator $\mathbf{s}_1 \cdot \mathbf{s}_2$ are equal to $-\frac{3}{4}$ and $\frac{1}{4}$ respectively for these two states. Hence, the eigenvalues of the operator $-J_1/2 - 2J_1\mathbf{s}_1 \cdot \mathbf{s}_2$, are just $\pm J_1$. Omitting the unimportant constants, we conclude that the two energies (7.11) are the eigenvalues of the operator $-2J_1\mathbf{s}_1 \cdot \mathbf{s}_2$. If one has many pairs of electrons associated with the sites of a lattice, one can describe them by the so-called Heisenberg quantum operator

$$H = -2J_1 \sum_{i,j} \mathbf{s}_i \cdot \mathbf{s}_j \, . \tag{7.13}$$

Thus, the antisymmetry of the wave function, a purely quantum property, allows us to describe the effective interaction between the particles in the form of the spin operators, even though the original Hamiltonian was independent of spin. For the two electrons in the helium atom, the exchange integral (7.12) is negative, and therefore, according to Eq. (7.13), the minimum energy corresponds to the antiparallel configuration of the electron spins. On the other hand, for ferromagnetic systems it is positive, and the parallel configuration of spins is preferable. We see that the Ising model (3.1) was a simplified version of the Heisenberg model (7.13), with scalar spins instead of vector operators.

One has to distinguish between the above "exchange interaction" (7.13) between spins and the interaction of the elementary magnetic moments μ connected with the spin $\hbar/2$ through the gyromagnetic ratio $e/(mc)$, $\mu = e\hbar/(2mc)$. While the former interaction is strong, being on the order of $1000\,K$, the latter is on the order of $1\,K$ and it is able to explain only the so-called dipole ferromagnetism. This makes clear what were the problems which P. Weiss had in 1907 (before the discovery of quantum mechanics) with his mean field theory described in Chapter 2.

7.3   Quantum Statistical Physics

Let us now consider some aspects of the statistical physics of quantum systems. We consider N non-interacting particles inside a vessel which is in contact with a thermal reservoir and a reservoir of particles at fixed volume. Such a system is described by the partition function of a grand canonical ensemble which has the form

$$Z = \sum_N \exp\left(\frac{N\mu}{kT}\right) \sum_n \exp\left(-\frac{E_{n,N}}{kT}\right) , \tag{7.14}$$

where $E_{n,N}$ are the energies of the different states n of a system containing N particles, and μ is the chemical potential. For such a system of non-interacting particles, the total number of particles present is $N = \sum_K n_K$ and the total energy $E = \sum_K \varepsilon_K n_K$, where n_K is the number of particles with energy ε_K. Then the free energy $\Omega = -kT \ln Z$ can be written in the form

$$\Omega = -kT \ln \left\{ \sum_{K,n_K} \exp[\beta n_K(\mu - \epsilon_K)] \right\} \equiv \sum_K \Omega_K, \qquad (7.15)$$

where as previously $\beta = 1/kT$.

For fermions, in accordance with the Pauli exclusion principle $n_K = 0$ or 1, and so

$$\Omega_K = -kT \ln\{1 + \exp[\beta(\mu - \epsilon_K)]\}. \qquad (7.16)$$

The situation is more complicated for bosons, where there are no restrictions on n_K. On summing the geometric series in the exponent of (7.15), one finds that

$$\Omega_K = -kT \ln \left\{ \frac{1}{[1 - \exp[\beta(\mu - \epsilon_K)]]} \right\}, \qquad (7.17)$$

provided that $\exp[\beta(\mu - \epsilon_K)] < 1$, i.e., that $\mu < \varepsilon_K$. Since for the state of lowest energy $\epsilon_K = 0$, this condition requires μ to be negative. The mean number of particles $< n_K >$ is defined from the relation

$$< n_K > = -\partial \Omega_K / \partial \mu, \qquad (7.18)$$

so that

$$< n_K > = \frac{1}{\exp[\beta(\epsilon_K - \mu)] \pm 1}, \qquad (7.19)$$

where the signs plus and minus are related to fermions and bosons, respectively.

We now consider the consequence of the requirement of negative μ for bosons. In order to do this, we need to derive the equation that determines μ. The volume of phase space available for a single state of a particle is $(2\pi\hbar)^{-1}$ [1], so that the number of states in phase space with momentum between p and $p + dp$ is

$$N(p)dp = \frac{4\pi gV p^2 dp}{(2\pi\hbar)^3}, \tag{7.20}$$

where V is the volume occupied by the system and $g = 2S + 1$ accounts for the different possible values of the spin. The chemical potential μ is defined from the conservation law of the number of particles,

$$N = \int N(p) < n(p) > dp. \tag{7.21}$$

On transforming to the variable of integration $\epsilon = p^2/2m$, substituting Eqs. (7.19) and (7.20) into Eq. (7.21) and making some simple transformations, one finds that

$$\frac{N}{V} = A \int \frac{\sqrt{\epsilon}d\epsilon}{\exp[\beta(\epsilon - \mu)] - 1}, \tag{7.22}$$

where A is some constant. With the change of variable $z = \beta\varepsilon$, this equation can be written in the form

$$\frac{N}{V} = A(kT)^{3/2} \int \frac{\sqrt{z}dz}{\exp(z - \beta\mu) - 1}. \tag{7.23}$$

Since N/V is constant, as the temperature decreases the integral must increase, and so the positive quantity $-\mu$ must decrease. However, μ vanishes not at $T = 0$, but at some finite temperature $T = T_c$ which is defined by the condition

$$\frac{N}{V} = A(kT_c)^{3/2} \int \frac{\sqrt{z}dz}{\exp(z) - 1}. \tag{7.24}$$

Up to now we have considered only particles in states with $\epsilon > 0$, and not allowed for the states with $\epsilon = 0$. In order to adjust Eqs. (7.23) and (7.24) so that they can be satisfied for $T < T_c$, one has to assume that Eq. (7.24) represents the number of particles in states with $\epsilon > 0$, and that all the remaining N_0 particles must be in the state with energy $\epsilon = 0$. Hence,

$$N_0 = N[1 - (T/T_c)^{3/2}], \qquad (7.25)$$

and a phase transition occurs at $T = T_c$, where the particles start filling up the state with $\epsilon = 0$. This quantum effect, which is called Bose–Einstein condensation, is the only known example of a phase transition in a system without interactions.

The question arises of how to observe Bose–Einstein condensation experimentally. In the above treatment it was assumed that the system considered was an ideal gas of bosons, with no interactions between the particles. However, because of inter-particle attractive interactions most gases solidify before the temperature becomes low enough for such quantum effects to be observable. A necessary condition for quantum effects to manifest themselves is that the thermal de Broglie wavelength λ of the particles be larger than the average distance r_0 between them. Since

$$\lambda = \frac{h}{p} = \frac{h}{Mv} = \frac{h}{M\sqrt{kT/M}}, \qquad (7.26)$$

where M is the mass of the particles, this condition can be written approximately as

$$\frac{h}{\sqrt{MkT}} > r_0. \qquad (7.27)$$

This requirement is quite difficult to satisfy, since a small value of r_0 implies a high concentration of particles, which favors solidification. In order to satisfy the inequality (7.27) one needs either very light particles or very low temperatures. For instance, the lightest simple boson is the He^4 atom, and for this Kapitza and Allen observed in 1937 superfluid behavior at atmospheric pressure and temperatures of around $2\,K$. Another possibility for satisfying the inequality (7.27) is to use very low temperatures for bosonic alkali metals, in which the inter-atomic interactions are comparatively weak, even though the mass of the atoms is not so small. Indeed, Bose–Einstein condensation has been observed recently in such systems at the incredibly low temperature of $10^{-7}\,K$. We will now consider first Bose–Einstein condensation in helium and the related phenomenon of superfluidity, and after that Bose–Einstein condensation in alkali metals.

7.4 Superfluidity

In this book, we will consider the simplest argument about super-fluidity, which was presented by Landau [1]. Let us imagine a fluid flowing with velocity $\mathbf{u}$ along a tube at zero temperature. In the rest frame of the fluid, the walls are moving with velocity $-\mathbf{u}$, and if there is friction the fluid near the wall must move at some velocity $\mathbf{v}$ between 0 and $-\mathbf{u}$, and so the energy of the fluid increases. If this is impossible, then the fluid will flow without friction, which is just the phenomenon of superfluidity. An increase of the fluid's energy can only occur if the walls transfer some energy to the liquid. According to quantum mechanics, the energy must be transferred in the form of quasi-particles with energy ϵ and momentum $\mathbf{p}$. In the original frame of reference, the energy obtained from the walls will be $\epsilon + \mathbf{p} \cdot \mathbf{v}$, and such an energy transfer is energetically favorable if this quantity is negative, $\epsilon + \mathbf{p} \cdot \mathbf{v} < 0$. The minimum value of the left hand side of this expression occurs when $\mathbf{p}$ is anti-parallel to $\mathbf{v}$, in which case the condition becomes $v > \epsilon/p$. Since only under this condition is an energy transfer possible, the crucial question that arises is the minimum value of the ratio ϵ/p. Two possibilities are shown in Fig. 7.1. For a free particle $\epsilon = p^2/(2m)$, and the minimum value of ϵ/p (which occurs at $p = 0$) is zero, so that such an energy transfer is always possible and the fluid flow will always be accompanied

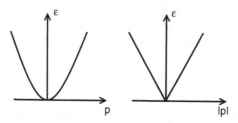

Fig. 7.1 The energy ϵ as a function of the momentum p (a) for free particles; (b) for particles with $\epsilon \sim |p|$.

by friction. However, in systems in which $\epsilon \sim |p|$ (as for phonons in solids), the minimum value of $d\epsilon/dp$ is c_0. It is found [39] that in liquid helium, the excitations have a dispersion law of the form $\epsilon = c_0|p|$, so that for velocities less than c_0 the fluid must flow without friction, i.e., it behaves as a superfluid. The above considerations apply to a system at zero temperature. For finite temperatures T that are less than T_c, some thermal excitations already exist, but new ones will not appear for velocities smaller than c_0. Therefore, for temperatures $0 < T < T_c$, ^{4}He behaves as a mixture of a normal fluid and a superfluid. Such a behavior was first observed in 1937 in liquid He^4, for which the value of c_0 is $60\,\text{cm/s}$, and was explained by Landau in the 1940's.

7.5   Bose–Einstein Condensation of Atoms

For over fifty years, liquid helium was the only "quantum" fluid showing superfluidity. Over the years, it became clear that the same effect exists, in principle, for atoms heavier than helium, as soon as, according to criterion (7.27), one reaches a temperature much lower than $2.2\,K$ which is the phase transition temperature for Bose–Einstein condensation in helium. The real breakthrough in low-temperature experimental physics occurred in the 1980s and 1990s, when researchers managed to attain the unbelievably low temperatures of a few hundred nano-kelvin, which is a few hundred billionths of a degree above absolute zero! (Note that the lowest temperature found in Nature, which comes from the Big Bang's residual radiation, is "only" $3\ K$.) At these ultra-low temperatures atoms are practically "stopped", having a velocity of a few centimeters per second. This achievement is especially impressive since it is impossible to control the behavior of neutral atoms by the use of the electric and magnetic fields commonly used for charged particles. The slowing down of atoms was achieved with the help of magneto-optical traps, using laser cooling by three mutually orthogonal laser beams and evaporative cooling by an inhomogeneous magnetic field [40]. It is no coincidence that after the 1997 Nobel prize "for the development of methods to cool and trap atoms with laser light" the turn came of the 2001 Nobel prize "for the achievement of Bose–Einstein condensation (BEC) in dilute gasses of alkali atoms...". A Bose–Einstein condensate is a collection of atoms in a single macroscopic quantum state described by one single wave function. This wave function of a "superatom" can now be calculated or observed as well as photographed! Many experimental groups have worked on this subject, and BEC has been observed in rubidium, potassium, lithium, cesium and spin-polarized hydrogen. Already in the first experiments on rubidium, the existence of a condensate was proved by measurements of the velocity distributions of the atoms, which after reaching the transition temperature changed from the Gaussian distribution appropriate to a classical gas to a narrow peak of velocities centred at zero velocity. Additional experiments include the resonance effect of two separate condensates, and the observation of a vortex lattice in a rotating condensate. The Bose–Einstein condensate is a new state of matter, and its study is very important for a full theoretical understanding of the macroscopic manifestation of quantum laws, which also appear in mesoscopic physics, a popular topic nowadays. Possible applications of BEC may include "atomic lasers" and different uses in nano-technology. This field is developing and changing rapidly, and we suggest that those reader interested in the subject should follow developments in the scientific periodicals. A simple presentation of the subject can be found in the Scientific American articles [41], and in Cornell's and Wieman's Nobel lectures [42].

7.6 Superconductivity

The experiments on and theories of superconductivity, and in particular of high temperature superconductivity, are a vast subject, which we will not even attempt to summarize in this book. However, since the phase transition from a normal to a superconducting state is of great theoretical and practical importance, it should not be omitted from a book on phase transitions. Accordingly, we present here a brief description of the main experimental facts and of the BCS (Bardeen–Cooper–Schrieffer) theory of type I superconductors, which is sufficient for appreciating the special features of the super-conductivity phase transition. After this, we will consider the main features of high temperature superconductors.

Superconductivity consists of the flow of electric currents in a material without any resistance, and so is a special case of super-fluidity, in which the electrons which are fermions create a bosonic "fluid". It was first observed by Kamerling-Ohnes in 1911, soon after he succeeded in liquefying helium and so obtaining very low temper-atures. He originally expected that his experiments would support Lord Kelvin's idea (the dominant one at that time) that at very low temperatures the electrons will be localized around ions, rather than be free, so that the conductivity will vanish [43]. However, in complete contrast to this, he found that as the temperatures T is lowered through a critical temperature T_c the electrical resistivity of mercury suddenly drops to zero (and stays there as the temperature is lowered more). Some other important properties of superconductor materials, discovered subsequently, include the following:

(1) Magnetic fields are expelled from type I superconductors. This absence of a magnetic field inside a superconductor, which does not follow from the infinite conductivity, is known as the Meissner effect. Incidentally, this effect is not found in type II supercon-ductors (the category to which belong all the high temperature superconductors discovered so far), which consist of coexisting superconducting and ordinary states.

(2) Although superconductivity is associated with the motion of elec-trons there is an isotope effect, so that for different isotopes of the same atom the critical temperature T_c is proportional to $1/\sqrt{M}$, where M is the mass of the atom isotope.

(3) Superconductivity can be destroyed by a sufficiently high mag-netic field or electrical current.

The standard theoretical explanation of superconductivity is the BCS theory. The main idea of this is that there is an attractive force between pairs of electrons, in spite of the Coulomb repulsion between them. Such a force is associated with the motion of the atoms in the solid, just as a ball rolling on the surface of a drum will depress the surface, and surrounding balls will be attracted into this depression. The most important property of the resulting Cooper pairs of electrons is that they have integer spin and so are bosons, since (as we saw above) Bose–Einstein condensation and superfluidity are only possible for such particles. In a metal, the electrons of interest are those originally in states close to the Fermi surface, i.e., with energies close to the Fermi energy ϵ_F, which are the ones that carry the electric current in normal metals. Since for electrons at the Fermi surface the minimum value of $d\epsilon/dp$ is finite, such a "bosonic" gas of electrons will exhibit superfluidity which — for charged particles — means superconductivity.

The only property which still has to be explained is the origin of these electron pairs. Here we follow the two-page paper of Cooper (from which originates the name Cooper pairs) published in 1956 [44]. We consider the wave function $\psi(\mathbf{r}_1, \mathbf{r}_2)$ of a pair of electrons, and assume that it depends only on the distance between them, $\psi = \psi(\mathbf{r}_1 - \mathbf{r}_2)$, while the potential also depends only on this distance. Schrodinger's equation for $\psi(\mathbf{r}_1, \mathbf{r}_2)$ can be written in the form

$$(-\hbar^2/2m)(\nabla_1^2 + \nabla_2^2)\psi + V(\mathbf{r}_1 - \mathbf{r}_2)\psi = (E + 2\epsilon_F)\psi, \qquad (7.28)$$

where the energy E is that of the pair relative to the energy at the Fermi surface, and must be negative for a pair to be formed. We take the Fourier transform of this equation, and write

$$\psi(\mathbf{r}_1, \mathbf{r}_2) = \sum_{\mathbf{K}} g_{\mathbf{K}} \exp[i\mathbf{K}\cdot(\mathbf{r}_1 - \mathbf{r}_2)], \qquad (7.29)$$

which corresponds to one electron of the pair having momentum $\hbar\mathbf{K}$ and the other momentum $-\hbar\mathbf{K}$. On substituting Eq. (7.29) into Eq. (7.28) we find that

$$\frac{\hbar^2 K^2}{m} g_{\mathbf{K}} + \sum_{\mathbf{K}'} V_{\mathbf{K}-\mathbf{K}'} g_{\mathbf{K}'} = (E + 2\epsilon_F) g_{\mathbf{K}}. \qquad (7.30)$$

Let us now assume a potential such that

$$V_{\mathbf{K}-\mathbf{K}'} = -V_0 \quad \text{for} \quad \epsilon_F < \epsilon_K, \epsilon_{K'} < \epsilon_F + \hbar\omega_D \qquad (7.31)$$

and $V_{\mathbf{K-K'}}$ is zero otherwise, where ω_D is the Debye frequency of the phonons in the metal, which is proportional to $1/\sqrt{M}$. Such a potential is suitable for describing the phonon induced coupling between pairs of electrons in a spherical shell just above the Fermi surface, and its microscopic derivation is not important for understanding the superconducting phase transition. It then follows from Eqs. (7.30) and (7.31) that

$$g_K = \frac{V_0 \sum_{\mathbf{K'}} g_{\mathbf{K'}}}{\hbar^2 K^2/m - E - 2\epsilon_F}. \tag{7.32}$$

The summation of both sides of Eq. (7.32) over K leads to

$$1 = \sum_K \frac{V_0}{\hbar^2 K^2/m - E - 2\epsilon_F}. \tag{7.33}$$

On going from a sum over K to an integral over the spherical shell in which V_0 is non-zero, and converting the variable of integration from K to

$$\xi = \hbar^2 K^2/(2m) - \epsilon_F, \tag{7.34}$$

we then find that

$$V_0 \int_0^{\hbar\omega_D} \frac{n(\xi)d\xi}{2\xi - E} = 1, \tag{7.35}$$

where $n(\xi)$ is the density of states in the spherical shell, which to a good approximation is the density of states $n(\epsilon_F)$ at the Fermi level. Hence,

$$\ln \frac{2\hbar\omega_D - E}{-E} = \frac{2}{V_0 n(\epsilon_F)}, \tag{7.36}$$

or

$$E \exp\left(\frac{2}{V_0 n(\epsilon_F)}\right) = E - 2\hbar\omega_D. \tag{7.37}$$

Since V_0 is assumed to be small and $n(\epsilon_F)$ finite, it follows that

$$\exp\left(\frac{2}{V_0 n(\epsilon_F)}\right) \gg 1. \tag{7.38}$$

Thus, finally,

$$E \approx -2\hbar\omega_D \exp\left(-\frac{2}{V_0 n(\epsilon_F)}\right) < 0 \qquad (7.39)$$

and so it it is energetically favorable for the two electrons to be bound together in a pair.

We now consider a number of other consequences of the Cooper formula (7.39):

(*a*) Firstly, we note that the binding energy E of a pair is proportional to ω_D and so to $1/\sqrt{M}$, so that, since one expects that $kT_c \sim |E|$, the critical temperature T_c will also be proportional to $1/\sqrt{M}$, which is just the isotope effect.

(*b*) Secondly, we note that V_0 enters the Cooper formula in the form $\exp\{-2/[V_0 n(\epsilon_F)]\}$, which has an essential singularity at $V_0 = 0$. Hence, although V_0 is very small, one cannot obtain Cooper pairs by applying any sort of perturbation theory which uses expansions in powers of V_0.

(*c*) Thirdly, in order to destroy the coupling (so that the electrons become independent and contribute to normal electric conductivity rather than to superconductivity), one has to supply them with an energy on the order of $|E|$ by increasing the temperature above T_c, or by an electric current.

(*d*) Fourthly, an essential element of the theory is that the electron pairs appear above the Fermi level, so that the relevant density of states is $n(\epsilon_F)$, and not some positive power of ϵ which would become zero at $\epsilon = 0$.

(*e*) Finally, the natural question arises as to why the Coulomb repulsion between the electrons does not destroy the Cooper pair. Leaving the full answer to the microscopic theory [39], the qualitative answer lies in the fact that the size of the Cooper pair is very large so that the Coulomb repulsion is small. In order to estimate the size a of the Cooper pair, we use the Heisenberg uncertainty principle, and write

$$a \approx \frac{h}{\delta p_F} \approx \frac{h p_F}{m \delta \epsilon_F} \approx \frac{h p_F}{m \hbar \omega_D} \approx \frac{h}{p_F} \frac{\epsilon_F}{\hbar \omega_D} \gg \frac{1}{k_F}. \qquad (7.40)$$

Since the mean distance between electrons is of order $1/k_F$, it follows that the electrons in a pair are very far apart, with many other electrons between them, which explains why their mutual Coulomb repulsion is so small.

7.7   Tunneling Supercurrent: Josephson Effect

So far we considered the superconductor as orderly array of electrons that form non-interacting Cooper pairs, which explains the phenomenon of superconductivity. New phenomena arise when two superconductors are placed together separated by a thin insulating or metallic barrier (SIS and SMS systems known as Josephson junctions). Although similar phenomena occur in other areas of physics (bipolarons [45], excitons as the combination of electron and hole [46], etc.) some unusual and exciting phenomena of supercurrent, appear in SIS and SMS, known as dc and ac Josephson effects [47]. A dc current flowing forever without any voltage applied and ac current induced by applying a constant voltage seem to defy our physical intuition. The Josephson effect is an example of a macroscopic quantum phenomenon which occurs in superconductors that represent macroscopic quantum systems. In fact, such an effect exists in any two macroscopic quantum system separated by a thin barrier.

Josephson based his discovery on the fact that each superconducting system, including SIS and SMS, is characterized by a two-dimensional order parameter — a wave function $\Psi = \Psi_0 \exp(i\phi)$. Since only $|\Psi|^2$ has physical meaning, two separate superconductors may have different phases ϕ. Due to its theoretical importance and many potential applications, Josephson was awarded the Nobel prize together with Anderson and Rowell, who checked these effects experimentally [48] and awarded patents on their work. The number of applications of the Josephson effects (voltmeters, magnetometers, precise measurements of $\hbar/e$) will certainly increase with time.

The rigorous description of the Josephson effects requires the framework of microscopic theory. For simplicity, in the next two subsections, we bring a phenomenological description, all of whose results are confirmed by the microscopic theory.

7.7.1 *dc Josephson effect*

We start with the well-known expression for the density of the superconducting current in the magnetic field $\mathbf{B} = curl\ A$,

$$j = \frac{e\hbar}{m} \left(\psi^* \frac{d\psi}{dx} - \psi \frac{d\psi^*}{dx} \right) - \frac{4e^2}{mc} \mid \psi \mid^2 A. \qquad (7.41)$$

If a system is placed in a bounded region of a space restricted by surface S, then

$$\oint \vec{j}\ \vec{dS} = 0. \qquad (7.42)$$

If the superconductor is located in the region $x > 0$ and the magnetic field B is parallel to the x-axis, one obtains from (7.41) and (7.42),

$$\oint j\ dS = \oint dS \left[\psi^* \left(\frac{d\psi}{dx} - \frac{2e}{\hbar c} A\psi \right) - \psi \left(\frac{d\psi^*}{dx} - \frac{2e}{\hbar c} A\psi^* \right) \right] = 0. \qquad (7.43)$$

It follows from Eq. (7.43) that the expressions in parentheses equal zero. This result has been obtained for a superconductor bordering a non-magnetic substance. In our case of SIS, where superconductor 1 borders superconductor 2 through a thin isolating film, so that Cooper pairs are able to penetrate this film, one obtains

$$\frac{d\psi_1}{dx} - \frac{2e}{\hbar c} A\psi_1 = a\psi_2; \quad \frac{d\psi_1^*}{dx} - \frac{2e}{\hbar c} A\psi_1^* = a\psi_2^*. \qquad (7.44)$$

Substituting (7.44) into (7.41) one obtains for tunneling supercurrent,

$$j_s = \frac{e\hbar a}{m} \left(\psi_1^* \psi_2 - \psi_1 \psi_2^* \right). \qquad (7.45)$$

Suppose now that both superconductors consist of the same material and differ only in phase, $\psi_1 = \psi_0 \exp(i\phi_1)$ and $\psi_2 = \psi_0 \exp(i\phi_2)$.

Then, Eq. (7.45) reduces to the first Josephson equation

$$j_s = j_0 \sin{(\phi_2 - \phi_1)}\,, \tag{7.46}$$

i.e., in the closed contour containing SIS, the tunneling supercurrent j_s flows forever without any voltage being applied.

7.7.2   *ac Josephson effect*

Taking into consideration the boundary conditions, as in Eq. (7.44), the Schrodinger equations for the superconductors 1 and 2 have the following form,

$$i\hbar\frac{\partial\psi_1}{\partial t} = E_1\psi_1 + a\psi_2; \quad i\hbar\frac{\partial\psi_2}{\partial t} = E_2\psi_2 + a\psi_1\,. \tag{7.47}$$

As in the previous subsection, assume that $\psi_1 = \psi_0 \exp(i\phi_1)$ and $\psi_2 = \psi_0 \exp(i\phi_2)$. Inserting these functions in Eq. (7.47) and equating the real parts in each equation, one gets

$$\hbar\frac{d\phi_1}{dt} = a \cos{(\phi_2 - \phi_1)} + E_1; \quad \hbar\frac{d\phi_2}{dt} = a \cos{(\phi_2 - \phi_1)} + E_2\,. \tag{7.48}$$

If a potential difference V is applied across the SIS contour, the difference in energies $E_2 - E_1$ is equal to $2eV$. Subtracting the two equations in (7.48) and using $E_2 - E_1 = 2eV$, leads to

$$\frac{d}{dt}(\phi_2 - \phi_1) = \frac{2e}{\hbar}V \tag{7.49}$$

or

$$\phi_2 - \phi_1 = (\phi_2 - \phi_1)_0 + \frac{2e}{\hbar}Vt\,. \tag{7.50}$$

Inserting (7.50) into (7.46) leads to the second Josephson equation,

$$j_s = j_0 \sin\left[(\phi_2 - \phi_1)_0 + \frac{2e}{\hbar}Vt\right]\,, \tag{7.51}$$

which shows that the ac tunneling supercurrent j_s is induced by the constant voltage V.

7.8 High Temperature Superconductors

More than ten thousand scientific articles on high-T_c superconductivity have appeared since its discovery. It is impossible, therefore, to describe here even the main lines of investigations, which also change rapidly, and so we restrict ourselves to a very general picture, and some recent (at the time of writing this book) results.

For more than seventy years after the discovery of superconductivity, it was found in hundreds of pure materials, binary alloys and more complicated materials. In some of these materials it was quite unexpected, for instance in MoC where neither component is a superconductor. However, the applications of these materials were quite restricted due to the low critical currents (so that large currents, which produce a large enough magnetic field, destroy the superconductivity) and to the low critical temperatures. The highest critical temperature, $23.1\,K$, which was achieved in Nb_3Ge, is only slightly above the boiling point of liquid hydrogen ($20.3\,K$ at a pressure of one atmosphere). Among the restricted number of applications of low-temperature superconductors, one of quite general interest is Magnetic Resonance Imaging, which uses superconductors as the source of strong magnetic fields and is used to obtain images of soft issues in the human body, similar to those of bones obtained by X-rays.

The new era which marked the birth of high-T_c superconductivity started in 1986 when Bednorz and Muller [49] found a material with a critical temperature of around $35\,K$. In the following year (!) they received a Nobel prize for this discovery, and following their discovery a large number of solid state laboratories started both an intensive study of different aspects of this phenomenon and a search for materials with even higher transition temperatures. Nowadays, a critical temperature of order $90\,K$ is obtained quite routinely. One important advantage of such high critical temperatures is that they are easily achieved by the liquefaction of nitrogen at $77\,K$. This is a much cheaper process than the liquefaction of helium which had to be used previously to obtain superconductivity. In addition to their high transition temperatures, the high-T_c superconductors also have high critical currents, which is very important for applications. The typical high-T_c superconductors (called in professional jargon "bisko" for BSCCO and "ibco" for YBCO) have the so-called perovskite-type structure with copper Cu, and oxygen O, located in well separated parallel planes of copper oxide, and atoms such as yttrium Y, bismuth B, strontium S, lanthanum La, or calcium Ca located between these planes. However, these materials have the disadvantage of being ceramics, which are difficult to manufacture in the form of wires as is required for most applications. This is one reason for the continuing intensive search for new superconducting materials which are easier to process mechanically and/or cheaper to produce, and not just for materials with higher critical temperatures or fields. One recent (April 2003) promising material in this class is magnesium diboride MgB_2, whose transition temperature of $39\,K$ is unexpectedly high for this type of material. Although this is well below the liquefaction temperature of nitrogen, it could lead to a family of similar materials with higher critical temperatures. This is why an extensive article in the March, 2003 issue of Physics Today was devoted to MgB_2 and also a special issue of a scientific journal [50] and the special April, 2003 Cambridge conference. Although the fragile ceramic structure of the standard high-T_c materials poses major problems for their practical applications, the Washington Post reported that American Superconductor Co. set a 1200-foot long electrical cable in the tunnel near Detroit which supplies electrical power for 14,000 customers. The same company supplied 18 miles of superconducting wires to Pirelli Co. The era of magnetically levitated high-speed trains and loss-free electrical lines is approaching....

The high-T_c superconductors raise many theoretical questions as well, the most important of which is whether they are described by the BCS theory which was so successful for describing the low temperature superconductors. Some additional ideas involve describing the Cooper pair as having the d-wave symmetry rather than spherical symmetric s-waves, as having a double-energy gap (for MgB_2, for example) rather than the simple gap described in the previous section, as having special "strips" connecting atoms, etc. In order to understand the avalanche of scientific publications we suggest studying first the simply written article in Contemporary Physics

[51], as well as different monographs specially dedicated to high-T_c superconductors.

7.9 Conclusion

The exchange interactions of fermions, which follow from their symmetry properties, is responsible, among other things, for the ordering of spins in ferromagnets described by the Heisenberg Hamiltonian. The statistical properties of bosons give rise to the phenomenon of Bose–Einstein condensation (BEC), in which particles accumulate in the state of zero energy. Superfluidity (the non-viscous motion of a system) is a macroscopic manifestation of BEC. A related phenomenon is superconductivity, in which electrons ordered in charged bosonic pairs also move without friction, or in other words without resistance. New developments in this area are connected with the discovery of BEC of atoms and superconductivity at high-T_C temperatures.

Chapter 8

Universality

In the previous chapters, we have seen that different models exhibiting phase transitions, such as the mean field model and the Ising model in two and three dimensions, have different sets of critical indices associated with the transition. Phase transitions are usually classified according to their sets of critical indices, such that systems with phase transitions having the same set of critical indices belong to the same universality class. This name is used because, as we show in this chapter, there are often a range of different systems belonging to the same universality class.

8.1 Heisenberg Ferromagnet and Related Models

As we saw in the last chapter, the exchange interaction between electrons leads to the Heisenberg quantum Hamiltonian

$$\hat{H} = -J \sum_{i,\, j} \mathbf{s}_i \cdot \mathbf{s}_j \,, \tag{8.1}$$

(with $2J_1 = J$) and the associated quantum partition function

$$Z = Tr\{\exp[-\hat{H}/(kT)]\} \,. \tag{8.2}$$

Since near a phase transition the correlation length ξ is very large, quantum effects are unimportant here, and so we can replace the quantum operator $\hat{H}$ by the Hamiltonian of the classical Heisenberg model,

$$H = -J \sum_{i,j} \mathbf{S}_i \cdot \mathbf{S}_j \,. \tag{8.3}$$

This model is isotropic, but in real systems crystal symmetry in conjunction with spin–orbit coupling can lead to anisotropy, with different values of J for the different principal symmetry axes of the crystal. Accordingly, we now consider the generalized Heisenberg Hamiltonian, for which the x-axis and y-axis are equivalent but differ from the z-axis, so that

$$H = -J' \sum_{i,j} S_i^z S_j^z - J \sum_{i,j} (S_i^x S_j^x + S_i^y S_j^y) \,. \tag{8.4}$$

The Ising model, for which $J = 0$, and the ordinary Heisenberg model, for which $J' = J$, are special cases of this generalized Hamiltonian. It is known that the critical indices for the Ising and Heisenberg models are different. For the Hamiltonian (8.4), the critical indices of the phase transition are found to be the same as for the Ising model whenever $J < J'$. Only for $J' = J$ does one obtains the same critical indices as for the Heisenberg model. A further special case, $J' = 0$, the so-called x–y model, will be considered separately later on in this chapter. Thus, the critical indices do not depend continuously on the interaction energies J and J', but rather jump from one "universality class", that of the Ising model for all $J < J'$, to the Heisenberg universality class for $J = J'$. In addition, for spatial dimensions $d \geq 4$ the critical indices belong to the mean field universality class which is different from both the Ising and the Heisenberg universality classes.

In general, the universality class of a system depends only on its spatial dimensions d, the number of components (dimensionality) D of the order parameter and on whether the range r_c of the interactions is finite or infinite. With regard to the dependence on d and D, we have already seen that for the Ising and Heisenberg models there is a difference between one, two and three dimensions, while the difference between the Ising and Heisenberg models is in the value of D, as the dimension of the order parameter for the Ising model is unity and for the Heisenberg model is three.

In order to obtain a better understanding of the role of the range of the interaction, let us consider the infinite-range Ising model of N spins described by the Hamiltonian

$$H = -\frac{J}{N} \sum_{i,\,j} S_i S_j \,, \tag{8.5}$$

where the summation is performed over all pairs of sites and not only over the nearest neighbors as it was in all previous formulae. The factor $1/N$ is needed since the energy has to be proportional to N, while the number of pairs in the sum is proportional to N^2. The double sum in Eq. (8.5) can be written as

$$\sum_{(i,j)} S_i S_j = \left(\sum_j S_j \right)^2 - N = S^2 - N \,, \tag{8.6}$$

where $S = \sum_j S_j$ is the total spin of the system. The substitution of Eq. (8.6) into Eq. (8.5) leads to the partition function (3.1) becoming

$$Z_N = Tr \left[\exp\left(-\beta J + \frac{\beta J}{N} S^2 \right) \right] = \exp\left(-\beta J \right) \sum_{S_j = \pm 1} \exp\left(\frac{\beta J}{N} S_j^2 \right). \tag{8.7}$$

The last exponent in Eq. (8.7) can be written as

$$\exp\left(a S_j^2 \right) = \frac{1}{\sqrt{4\pi a}} \int_{-\infty}^{\infty} du \exp\left(-\frac{u^2}{4a} + u S_j \right), \tag{8.8}$$

so that, on inserting Eq. (8.8) into Eq. (8.7), we find that

$$\mathbf{Z}_N = \sqrt{\frac{N}{4\pi\beta J}} \exp\left(-\beta J \right) \sum_{S_j = \pm 1} \int_{-\infty}^{\infty} du \exp\left(-\frac{Nu^2}{4\beta J} + u S_j \right)$$

$$= \sqrt{\frac{N}{4\pi\beta J}} \exp\left(-\beta J \right) \int_{-\infty}^{\infty} du \exp\left(-\frac{Nu^2}{4\beta J} \right) [2\cosh(u)]^N \,. \tag{8.9}$$

As $N \to \infty$ the Gaussian in Eq. (8.9) has a maximum that resembles a delta-function, and one can use the well-known method of steepest descents to show that the peak of the integrand is centered at

$$u^* = 2\beta J \tanh(u^*).\tag{8.10}$$

This equation is identical to Eq. (2.2), with u^* in place of M/M_0. It follows that the infinite-range Ising model (8.5) belongs to the mean field universality class. The above approximate calculation is appropriate for any dimension as long as each spin interacts with each other spin. The same mean field result follows from the exact solution of the one-dimensional model [52], in which the interaction $f(r)$ is very weak and decays exponentially with distance,

$$f(r) = -\gamma \exp\left(\frac{-r}{1/\gamma}\right) \quad \text{with } \gamma \to 0.\tag{8.11}$$

The reason that both systems with infinite-range interactions and systems in more than four space dimensions belong to the same universality class, that of mean field systems, is that such a large number of interactions affect each spin that their contribution can be well approximated by their mean value. Hence, systems with identical parameters d, D and the same range of interaction (i.e., finite or infinite) belong to the same universality class while other parameters do not influence the critical behavior. We now bring some examples.

(1) The Potts model [16], which has the same Hamiltonian as the Ising model but a spin S_j that can have the $(2S + 1)$ possible values $S, S-1, \ldots, -S$ for arbitrary integer or half-integer values of S, belongs to the same universality class as the Ising model, for which $S_j = \pm 1/2$.

(2) The critical behavior does not depend on the interaction energy and on the local structure of a system. For instance, it does not make any difference whether the set of spins considered are located on a simple cubic lattice, a body-centered one, a face-centered one, or a diamond-type lattice, even though the number of nearest neighbors of a site vary from four for the diamond lattice to twelve for the face-centered cubic lattice.

8.2   Many-Spin Interactions

We now describe briefly some more model systems. So far we have
considered different forms of pair interactions between spins. Another
group of models takes into account many-spin interactions described
by a Hamiltonian of the form

$$H = -J_2 \sum_{(i,j)} S_i S_j - J_3 \sum_{(i,j,k)} S_i S_j S_k - J_4 \sum_{(i,j,k,l)} S_i S_j S_k S_j - H \sum_i S_i .$$

(8.12)

Let us quote some of the known solutions for Hamiltonian (8.12):

a) $J_3 = J_4 = H = 0$, i.e., only pair interactions are taken into
account. The exact Onsager solution exists in two dimensions. For
three dimensions, the critical indices have been found by numerical
methods.

b) Baxter found [53] the exact solution for a two-dimensional
lattice composed of two mutually penetrating sublattices. The pair
interactions in each of the sublattice and quartet interactions of four
nearest neighbors were taken into account. It was shown by Wu [54]
that Baxter's case corresponds to $J_3 = H = 0$ in equation (8.12) and
to the inclusion of nearest and a few next pair interactions as well
as the quartet interactions of four nearest neighbors. The surprising
result of Baxter is the continuous dependence of the critical indices
on the interaction energy J_4. Such non-universal behavior turns out
to be an unpleasant surprise in the orderly picture of the theory of
phase transitions. Kadanoff and Wegner [55] explained which specific properties of the Baxter model is responsible for this unusual
property.

There are several other cases in which non-universality can
appear. For instance, let us consider the Union Jack lattice shown in
Fig. 8.1, which is described by the following Hamiltonian

$$H = -J_1 S_0 (S_1 + S_2 + S_3 + S_4)$$

$$- \frac{J_2}{2}(S_1 S_2 + S_2 S_3 + S_3 S_4 + S_4 S_1) - J_3(S_1 S_3 + S_2 S_4) . \quad (8.13)$$

The factor $1/2$ in the second term of Eq. (8.13), like that in Eq.
(6.20), takes into account the fact that each of the corresponding

interactions appears twice, in two adjacent cells. The particular cases, $J_1 = J_3 = 0$ and $J_2 = J_3 = 0$, correspond to the nearest-neighbor Ising model. Moreover, when $J_3 = 0$ and both J_1 and J_2 are non-zero, the exact solution [56] again exhibits the Ising type critical behavior. However, it has been shown [57] that the inclusion of an additional next-nearest-neighbor interaction J_3, the value of which is determined by J_1 and J_2, leads to non-universal eight-vortex type behavior.

Although the last example seems to be quite artificial, new types of phase transitions may appear for models with competing two-spin interactions. So far we have considered only one type of interaction in any given system, ferromagnetic for $J > 0$, and antiferromagnetic for $J < 0$. However, the possibility exists that in some systems the interactions are competing. For instance the interaction between nearest neighbors can be ferromagnetic and that between next nearest neighbors antiferromagnetic. In the next chapter we will consider the so-called spin glasses, which are characterized by such competing interactions.

c) Baxter and Wu [58] obtained the exact solution for the triangle lattice with triplet interactions ($J_1 = J_4 = H = 0$). It turned out that the critical indices are different from the Onsager result.

d) Imbro and Hemmer [59] investigated the critical behavior of the Hamiltonian (8.12) with $J_4 = H = 0$. It was found that only at $J_2 = 0$, i.e., in case c) are the critical indices different from Onsager

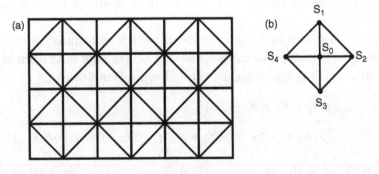

Fig. 8.1   (a) The Union Jack lattice; (b) Elementary cell.

values, whereas the critical indices have their Onsager values for all $J_2 \neq 0$, no matter what the relation between J_2 and J_3 is.

e) The recent great achievements of the theory of the phase transitions are connected with the renormalization group methods. The results for the critical indices is that they coincide with the Onsager critical indices to high accuracy.

Thus we can conclude from the cases d) and e) that the large fluctuations near the critical point cause an effective increase of pair interactions determining the critical indices and the many-body interactions are found to be inessential. In case c) when the pair interactions are absent, the critical indices are determined by triplet interactions. Except for the case b) the critical indices are not dependent on the interaction energy.

Additional analysis can be performed by the transformation of equation (8.12) from the discrete variables S_i to $\sigma(x)$ depending on the continuous space coordinate x [60]. Writing the fluctuation term as $\frac{1}{2}|\nabla \sigma(x)|^2$ we obtain instead of (8.12),

$$H = \int \left[u_2\sigma^2(x) + u_3\sigma^3(x) + u_4\sigma^4(x) - H\sigma(x) + \frac{1}{2}|\nabla \sigma(x)|^2 \right] d^dx.$$
(8.14)

Based on the universality hypothesis the original discrete model (8.12) and its continuous version (8.14) are taken to have the same critical behavior. In the framework of the continuous model (8.14) it is possible to perform the change of variable $\sigma(x) \to \sigma(x) + M$ in order to eliminate the triplet interactions in (8.14) [61], [62]. Finally we obtain the Hamiltonian (8.14) with $u_3 = 0$ and reorganized energy of pair interactions u_3 and "effective" magnetic field H:

$$u_2 \to \frac{1}{2}\left(u_2 - \frac{u_3^2}{u_4}\right); \qquad H \to \frac{u_3}{4u_4}\left(u_2 - \frac{u_3^2}{2u_4}\right)$$
(8.15)

(We consider here the case of zero magnetic field). Thus, accepting the validity of transformation from (8.12) to (8.14) we obtain for the Hamiltonian (8.12) the same Onsager critical indices as for case $J_3 = 0$ (except the case $J_2 = J_4 = 0$).

According to (8.15) there are three singularity points in this case:

$$\tilde{u}_3 = 0; \qquad \tilde{u}_2 = -2|\tilde{u}_{2,C}| \qquad \text{— the Onsager critical indices},$$

$$\tilde{u}_2 = 2|\tilde{u}_{2,C}|; \quad \tilde{u}_3^2 = 4|\tilde{u}_{2,C}|\tilde{u}_4 \qquad \text{— the Onsager critical indices},$$

$$J_2 = J_4 = 0; \quad \hat{J}_3 = \tilde{u}_{3,C} \qquad \text{— the Baxter-Wu critical indices},$$

$$(8.16)$$

where $\tilde{u}_i = u_i/T$, $\hat{J}_i = J_j/T$ and the dimensionless values $\tilde{u}_{2,C} = u_{2,C}/T$, $\tilde{u}_{3,C} = u_{3,C}/T$ determine the critical temperatures in the above considered models e) and c), respectively.

Let us investigate now the thermodynamic functions near each of the critical points (8.16). We can rewrite the Hamiltonians (8.12) and (8.14) near points (8.16) as a sum of the main term H_0 and the correction H_1

$$H_0 = J_2 \sum_{i,\,j} S_i S_j + J_4 \sum_{i,j,k,l} S_i S_j S_k S_l \,;$$

$$H_1 = J_3 \sum_{i,j,k} S_i S_j S_k \,,$$

$$H_0 = \int \left[u_2 \sigma^2(x) + u_4 \sigma^4(x) + \frac{1}{2} |\nabla \sigma(x)|^2 \right] d^d x \,;$$

$$(8.17)$$

$$H_1 = -h \int \sigma(x) \, d^d x \,,$$

$$H_0 = \sum_{i,j,k} S_i S_j S_k \,;$$

$$H_1 = J_2 \sum_{i,j} S_i S_j + J_4 \sum_{i,j,k,l} S_i S_j S_k S_l \,.$$

According to operator algebra [63] any function of spin variables can be rewritten as a linear combination of some basic operators. Near the critical points only two of them, the operator of magnetic moment M and of the energy E, determine the singularities of thermodynamic functions. Therefore, the operator H_1 in equations (8.17) can be presented in the following form

$$H_1 = a_1 M + a_2 E, \qquad (8.18)$$

where a_1 and a_2 are some coefficients. The second term in (8.18) causes only a renormalization of the critical temperature and will be omitted here. Finally

$$H_1 = -\hat{h}\, M, \qquad (8.19)$$

where the coefficient $\hat{h}$ acts as an effective magnetic field.

Comparing (8.17) and (8.19) we find the expressions for the effective fields in all three cases in Eq. (8.17)

$$\hat{h}_1 \sim J_3; \qquad \hat{h}_2 \sim h; \qquad \hat{h}_3 \sim b_1 J_2 + b_2 J_4, \qquad (8.20)$$

where b_1 and b_2 are some coefficients. In the last equality in (8.20) the property [64] that the operator $S_i S_j$ is proportional to the magnetic moment operator was used.

By scaling theory [35] the singular part of the thermodynamic potential can be written in the following form,

$$\Phi = \tau^{2-\alpha} f\left(\tau / \hat{h}^{1/\delta}\right). \qquad (8.21)$$

Using Eqs. (8.21) and (8.20), one can easily perform the full thermodynamical analysis [60].

Finally we conclude that the critical behavior of Ising systems is determined by pair rather than many-body interactions, except for the very special Baxter case (example b).

8.3 Gaussian and Spherical Models

The discrete Ising models with two-spin interactions can be generalized by other means. Two such models, the Gaussian model and the spherical model, were introduced by Mark Kac, to join "the small and select company of models in statistical mechanics which can be rigorously discussed" [65]. In both of these models instead of the spins S_j

having a finite number of discrete values ($\pm\frac{1}{2}$ in Ising model), they are continuous variables that can have any real value ($-\infty < S_j < \infty$). The similarity of these models to the Ising model is as follows. For the Gaussian model, it lies in the additional requirements that for each spin

$$\langle S_j^2 \rangle = \frac{1}{4} \,. \tag{8.22}$$

The condition (8.22) is a "relaxed" version of the "strong" condition $S_j^2 = \frac{1}{4}$ in the Ising model. The spherical model is closer to the Ising model than the Gaussian model in that not only must Eq. (8.22) be satisfied, but there is also the overall restriction

$$\sum_{j=1}^{N} S_j = \frac{N}{2} \,. \tag{8.23}$$

Both conditions (8.22) and (8.23) are obviously fulfilled in the Ising model.

Let us start with the Gaussian model. A Gaussian distribution of the spin variables is assumed, with $\langle S_j \rangle = 0$ and $\langle S_j^2 \rangle = \frac{1}{4}$, so that the probability $P(S_j)\,dS_j$ of finding a value S_j of the spin between S_j and $S_j + dS_j$ is given by

$$P(S_j)\,dS_j = \frac{1}{\sqrt{2\pi}} \exp\left(-\frac{1}{2}S_j^2\right) dS_j \,. \tag{8.24}$$

Then, the partition function Z has the following form

$$Z = \left(\frac{1}{2\pi}\right)^{\frac{N}{2}} \int_{-\infty}^{\infty} \cdots \int_{-\infty}^{\infty} dS_1 \cdots dS_N \exp\left(-\frac{1}{2}\sum_j S_j^2 + \frac{J}{kT}\sum_{i,\,j} S_i S_j\right) .$$
$$\tag{8.25}$$

The integration in Eq. (8.25) can be easily performed using Eq. (8.6), and the result leads to the conclusion that the Gaussian model does not show a phase transition in any number of spatial dimensions. This result is not so surprising, since after transforming the integral in Eq. (8.25) to "normal coordinates" the partition function does not contain any interaction between these coordinates, and in such a system without interactions there are no phase transitions in any number of spatial dimensions.

For the spherical model, the integration in Eq. (8.25) has to be performed under the additional constraint (8.23) which can be written in the form of the delta function, $\delta(\sum_{j=1}^{N} S_j^2 - N)$. If the integral representation of the delta function

$$\delta\left(\sum_j S_j^2 - N\right) = (2\pi)^{-1} \int_{-\infty}^{\infty} \exp\left[iu\left(\sum_j S_j^2 - N\right)\right] du \quad (8.26)$$

is substituted in Eq. (8.25), the following form of the partition function is obtained for the spherical model:

$$Z = \left(\frac{1}{2\pi}\right)^{\frac{N}{2}+1} \int_{-\infty}^{\infty} du \int_{-\infty}^{\infty} \cdots \int_{-\infty}^{\infty} dS_1 \cdots dS_N$$

$$* \exp\left[iu\left(\sum_i S_i^2 - N\right) - \sum_j \frac{S_j^2}{2} + \frac{J}{\kappa T} \sum_{i\,j} S_i S_j\right]. \quad (8.27)$$

Once again the transformation (8.6) can be used to perform integrations over S_j, just as was done for the Gaussian model, and the remaining integral over u, for the large values of N in which we are interested, can be carried out by the method of steepest descent. It turns out that a phase transition in the spherical model occurs for three and higher dimensional systems with critical indices different from those of the Ising model [66].

A geometric picture helps to clarify the relation between the Ising, Gaussian and spherical models [67]. If one considers each S_j as a component of an N-dimensional vector, then the Gaussian model corresponds to integration throughout space, the spherical model to integration over a spherical shell of radius $\sqrt{N}$, and the Ising model to summation over the corners of the unit hypercube inscribed within the unit sphere.

8.4 The x–y Model

We now turn to a model with properties very different from the above ones, namely the x–y model, for which the Hamiltonian is

$$H = -J \sum_{i,j} (S_i^x S_j^x + S_i^y S_j^y) \,, \qquad (8.28)$$

which corresponds to the generalized Heisenberg Hamiltonian of equation (8.4) with $J' = 0$. This model, with a two-component order parameter, is equivalent to one with $\psi = \psi_0 e^{i\theta}$, since one can write $S_i^x = \psi_0 \cos(\theta_i)$ and $S_i^y = \psi_0 \sin(\theta_i)$, where θ_i is the polar angle. Due to the common multiplier J in Eq. (8.28), we can put $\psi_0 = 1$. Then

$$H = -J \sum_{i,j} \left(\cos\theta_i \cos\theta_j + \sin\theta_i \sin\theta_j \right)$$

$$= -J \sum_{i,j} \cos(\theta_i - \theta_j) \,. \qquad (8.29)$$

For low temperatures the argument of the cosine is small, and so we can expand it in terms of $\theta_i - \theta_j$. Then, apart from a constant term, the Hamiltonian (8.29) takes the following form

$$H = \frac{J}{2} \sum_{i,j} (\theta_i - \theta_j)^2 \,. \qquad (8.30)$$

Since the Hamiltonian (8.30) is a function of the polar angle θ, it is invariant under the transformation

$$\theta_i \to \theta_i + 2\pi \,. \qquad (8.31)$$

We now go from the lattice model, in which the angle θ_j is defined on the sites j of a crystal lattice, to the continuous argument $\mathbf{r}$, $\theta = \theta(\mathbf{r})$. Then, the Hamiltonian (8.30) becomes

$$H = \frac{J}{2} \int |\nabla \theta(\mathbf{r})|^2 \, d^d r \,. \qquad (8.32)$$

Let us consider the correlation function $g(r_1 - r_2)$ of the phase of the order parameter

$$g(\mathbf{r}_1 - \mathbf{r}_2) = \int \exp\{i[\theta(\mathbf{r}_1) - \theta(\mathbf{r}_2)]\} \exp\left[-\frac{J \int |\nabla \theta(\mathbf{r})|^2 d^d r}{2kT}\right] d\theta.$$

(8.33)

Using the (real) Fourier transform of $\theta(\mathbf{r})$:

$$\theta(\mathbf{r}) = \sum_{\mathbf{K}} [\phi_{\mathbf{K}} \cos(\mathbf{K} \cdot \mathbf{r}) + \zeta_{\mathbf{K}} \sin(\mathbf{K} \cdot \mathbf{r})],$$

(8.34)

one can rewrite Eq. (8.33) as

$$g(R) = \iint d\phi_{\mathbf{K}} d\zeta_{\mathbf{K}} \exp \sum_{\mathbf{K}} \left[A_{\mathbf{K}}\phi_K + B_{\mathbf{K}}\zeta_K - \frac{JK^2}{2kT}\left(\phi_{\mathbf{K}}^2 + \zeta_{\mathbf{K}}^2\right)\right],$$

(8.35)

where $\mathbf{R} = \mathbf{r}_1 - \mathbf{r}_2$ and

$$A_{\mathbf{K}} = \cos(\mathbf{K} \cdot \mathbf{r}_1) - \cos(\mathbf{K} \cdot \mathbf{r}_2), \qquad B_{\mathbf{K}} = \sin(\mathbf{K} \cdot \mathbf{r}_1) - \sin(\mathbf{K} \cdot \mathbf{r}_2).$$

(8.36)

On going from a sum over $\mathbf{K}$ in the exponent of Eq. (8.35) to an integral in d dimensions, and completing the squares in ϕ_K and $\zeta_{\mathbf{K}}$, one finds finally that

$$g(R) = \exp\left\{-\frac{kT}{2J} \int \frac{d^d K}{K^2} \sin^2\left(\frac{KR}{2}\right)\right\}.$$

(8.37)

It follows that for the different spatial dimensions d as $R \to \infty$:

$$g(R) \sim \exp\left(-\frac{R}{\text{Const}}\right), \qquad d = 1,$$

$$g(R) \sim \text{Const} \cdot R^{\frac{-kT}{2\pi J}}, \qquad d = 2, \qquad (8.38)$$

$$g(R) \sim \text{Const} \cdot \exp\left(\frac{\text{Const}}{R}\right), \qquad d = 3.$$

In general, when $R \to \infty$ the correlation function $g(R)$ for a disordered system will decay exponentially, while in a completely ordered state it will be constant, showing the presence of long range correlations. It follows from Eq. (8.38) that there is no phase transition in a one-dimensional system while in three dimensions the system has long-range correlations. However, for $d = 2$ the correlation function has a power law decay, which corresponds neither to order nor to disorder, but rather to what is called quasi-order. Our calculation represents a special case of the general Mermin–Wagner theorem about the absence of a phase transition in a one-dimensional and a two-dimensional Heisenberg model [68]. In the last chapter we will consider such a power-law dependence as a fingerprint of self-organized criticality in non-equilibrium systems. For the present, we examine in more detail the two-dimensional x–y model, and compare it with the Landau mean field theory.

The order parameter η in the Landau theory can be written as

$$\eta = \Psi_0 \exp(i\theta_0). \tag{8.39}$$

In the standard (with no fluctuations) Landau theory, Eq. (4.6), a phase transition occurs at a temperature T_L, and the amplitude of the order parameter Ψ_0 in the ordered phase is equal to $\sqrt{-A/(2B)}$, while the phase θ_0 is arbitrary. This theory ignores fluctuations, and so gives too low a value for the entropy S. According to Eq. (3.17), a decrease of entropy leads to an increase of the phase transition temperature. Thus, the critical temperature T_L predicted by the Landau theory will be higher than that predicted by a theory that takes fluctuations into account. In order to include fluctuations in the Landau theory, we proceed as in Chapter 4, using the amplitude and the phase of the order parameter. We replace M in Eq. (4.20) by

$$\psi(\mathbf{r}) = \psi_0 + \psi_1(\mathbf{r})e^{i\theta(\mathbf{r})}, \tag{8.40}$$

where $|\psi_0|^2 = -\frac{A}{2B}$ as predicted by the Landau theory. Then

$$G = G_0 - \int \left[2A\,|\psi_1(r)|^2 - |\nabla\psi_1(r)|^2 \right] d^d r + \psi_0^2 \int |\nabla\theta(r)|^2) d^d \mathbf{r} + \cdots. \tag{8.41}$$

Just as in the calculations of Chapter 4, we use the Gaussian approximation for fluctuations, and leave only quadratic terms in Eq. (8.41). The angular part of the Hamiltonian (8.41) is identical to the x–y model (8.32). For high temperatures, $T > T_L$, the system is in a disordered state with exponentially decaying correlations. At $T = T_L$, a non-zero amplitude of the order parameter appears, but there are still fluctuations of the phase of the order parameter. Another phase transition, the so-called Kosterlitz–Thouless phase transition occurs at some temperature T_{KT} smaller than T_L, $T_{KT} < T_L$. This phase transition is connected with the unbinding of vortices, which are bound in pairs for $T < T_{KT}$, and become free for $T_{KT} < T < T_L$, as discussed below.

8.5 Vortices

The equilibrium states corresponding to the x–y Hamiltonian (8.32) can be obtained from the Lagrange equation

$$\frac{\delta H}{\delta \theta} - \nabla \frac{\delta H}{\delta(\nabla \theta)} = 0, \tag{8.42}$$

which for the x–y model reduces to

$$\nabla^2 \theta = 0. \tag{8.43}$$

For the two-dimensional case in which we are interested, the solution of Eq. (8.43) has the form

$$\nabla \theta = \frac{1}{r}, \qquad \theta = \ln(r). \tag{8.44}$$

The non-physical singularity of the function $\theta(r)$ at $r = 0$ can be removed by cutting out a small circle of radius a near the origin. This means that one includes a special type of fluctuation with zero order parameter in the small region surrounding $r = 0$. Such excitations around the equilibrium state are called topological charges or vortices, and the region $0 < r < a$ is called the core of the vortex. Vortices are characterized by their winding numbers, which are defined as follows. Consider the integral $\oint \nabla \theta \cdot d\mathbf{r}$ around any closed path enclosing the origin $r = 0$. Each circuit around the origin will, according to the symmetry property (8.31), bring us back to the original system with an increase of $2n\pi$ in the value of θ,

$$\oint \nabla\theta \cdot d\mathbf{r} = 2\pi n . \tag{8.45}$$

Here n, which is called the winding number of the vortex, is a positive (for vortices) or negative (for anti-vortices) integer, depending on whether the rotation is clockwise or anti-clockwise.

To answer the question of whether vortices will appear as result of thermal fluctuations at temperature T, one has to calculate the change in the free energy $F = E - TS$ associated with the appearance of vortices. It follows from Eq. (8.32) and (8.44) that the energy required to produce a vortex is

$$E = \frac{J}{2} \int_a^L |\nabla\theta|^2 d^2r = \pi J \ln\left(\frac{L}{a}\right) , \tag{8.46}$$

where L is a linear dimension of the system. On the other hand, since a vortex can appear everywhere in the system, the entropy of a vortex is

$$S = k \ln\left(\frac{\pi L^2}{\pi a^2}\right) = 2k \ln\left(\frac{L}{a}\right) . \tag{8.47}$$

Therefore if $T > T_{KT} = \pi J/(2k)$, the free energy $F = E - TS$ is negative, so that the appearance of a single vortex is energetically favorable. As we will see in the next section, for temperatures $T < T_{KT}$, the interaction between vortices results in the binding of vortex–antivortex pairs.

Incidentally, there is another example of a topological charge which is quite well known, though not necessarily by this name. This is the case of domain walls in ferromagnetic materials, which separate regions having opposite magnetizations. The number of domain walls increases as the critical temperature is approached from below, and the ferromagnetic–paramagnetic phase transition can be regarded as a topological one, originating from the proliferation of domain walls.

8.6   Interactions Between Vortices

The final point that we want to consider in connection with the x–y model is the interaction between vortices, and how they annihilate each other when the temperature is lowered below the Kosterlitz–Thouless temperature T_{KT}. Using Stokes theorem, one can rewrite Eq. (8.45) as

$$\oint \boldsymbol{\nabla}\theta \cdot d\mathbf{r} = \int (\nabla \times \boldsymbol{\nabla}\theta) \cdot d\mathbf{S} = 2\pi n \,. \qquad (8.48)$$

The second integral in Eq. (8.48) is identically zero unless there are singularities at isolated points $r = r_i$ within the contour, in which case

$$(\nabla \times \boldsymbol{\nabla}\theta) = 2\pi \mathbf{e}_z \sum_i n_i \delta(\mathbf{r} - \mathbf{r}_i) \equiv 2\pi \mathbf{n}(\mathbf{r}) \,, \qquad (8.49)$$

where $\mathbf{e}_z$ is the unit vector perpendicular to the xy-plane. From these two equations one finds that

$$\int \mathbf{n}(\mathbf{r}) \cdot d\mathbf{S} = n \,. \qquad (8.50)$$

Let us define the velocity $\mathbf{v}$ of the phase θ to be its spatial gradient $\nabla\theta$, $\mathbf{v} \equiv \boldsymbol{\nabla}\theta$. Since any vector $\mathbf{v}$ can be expressed as the sum of a longitudinal part $\mathbf{v}_l$ and a transverse part $\mathbf{v}_t$,

$$\mathbf{v} = \mathbf{v}_l + \mathbf{v}_t \,, \qquad (8.51)$$

where $\nabla \cdot \mathbf{v}_t = 0$ and $\nabla \times \mathbf{v}_l = 0$, we assume such a division for $\mathbf{v}$, and so consider only $\mathbf{v}_t$, for which $\nabla \times \mathbf{v}_t = 2\pi \mathbf{n}(\mathbf{r})$ and $\nabla \cdot \mathbf{v}_t = 0$, so that

$$\nabla \times (\nabla \times \mathbf{v}_t) = -\nabla^2 \mathbf{v}_t = 2\pi \nabla \times \mathbf{n}(\mathbf{r}) \,. \qquad (8.52)$$

We solve this equation using the Green's function $G(\mathbf{r} - \mathbf{r}')$ for the operator ∇^2, i.e., the solution of

$$\nabla^2 G(\mathbf{r} - \mathbf{r}') = -\delta(\mathbf{r} - \mathbf{r}') \,, \qquad (8.53)$$

which for a two dimensional system is given by

$$G(\mathbf{r} - \mathbf{r}') = \frac{1}{2\pi} \ln \left(\frac{|\mathbf{r} - \mathbf{r}'|}{a} \right).$$

(8.54)

Then, the solution of Eq. (8.52) can be written as

$$v_t(r) = -2\pi \int d^2 r' G(\mathbf{r} - \mathbf{r}') \left[\nabla \times \mathbf{n}(\mathbf{r}') \right]$$

$$= -2\pi \int d^2 r' \mathbf{n}(\mathbf{r}') \times \nabla G(\mathbf{r} - \mathbf{r}')$$

(8.55)

by using the standard formula for $\nabla \times (G\mathbf{n})$ and Stokes' theorem. The energy of the system of vortices, U, is obtained by the substitution of Eq. (8.55) into Eq. (8.32), which gives

$$U = 2\pi^2 J \int d\mathbf{r} d\mathbf{r}' d\mathbf{r}'' [\nabla G(\mathbf{r} - \mathbf{r}') \times \mathbf{n}(\mathbf{r}')] \cdot [\nabla G(\mathbf{r} - \mathbf{r}'') \times \mathbf{n}(\mathbf{r}'')].$$

(8.56)

On performing the partial integration over $\mathbf{r}$ and using Eq. (8.53) and (8.54), one finds that

$$U = \pi J \int dr' dr'' \ln \left(\frac{|\mathbf{r}' - \mathbf{r}''|}{a} \right) n(\mathbf{r}') n(\mathbf{r}'').$$

(8.57)

This equation, which describes the energy U of interaction between two vortices, can be compared to the energy E of an individual vortex given by Eq. (8.46). While E is positive and diverges logarithmically with the size L of the system, the interaction energy U is negative for the vortex–antivortex pair. Thus, if single vortices and antivortices were to appear at low temperatures, $T < T_{KT}$, they would immediately bind together. As we have seen earlier, only for temperatures $T > T_{KT}$ does the appearance of individual vortex becomes energetically favorable. Without going into any further details, we only note [69] that for the topological Kosterlitz–Thouless phase transition the correlation length ξ is always infinite below T_{KT}, while above T_{KT}

it diverges according to the non-trivial law,

$$\xi \sim \exp(\text{Const}/\sqrt{T - T_{KT}}).\tag{8.58}$$

Note the clear analogy between the behavior of vortices and that of the two-dimensional Coulomb gas with charges n and their interaction described by Eq. (8.57). In electrostatic language, the Kosterlitz–Thouless transition is analogous to unbinding the dipoles into single charges (a sort of insulator–metal transition).

8.7 Vortices in Superfluids and Superconductors

The x–y model is isomorphic to the superfluid and superconducting transitions, where the two-dimensional order parameter describes the wave functions of the Bose–Einstein condensate and of the Cooper pairs, respectively, while the interaction energy J has to be replaced by the superfluid density or by the electrical resistance of the two-dimensional films. In a three-dimensional superconductor, the presence of an external magnetic field will make the system effectively two-dimensional in the plane perpendicular to the field. The discovery of the high temperature superconductors stimulated a huge interest in the theory of vortices, since these materials are type II superconductors where the magnetic fields enters the superconducting materials in the form of vortices. We refer the reader to the current scientific literature for further details. As a starting point, one can use the comprehensive reviews of the theory of vortices in superfluids and superconductors [70], [71].

8.8 Conclusion

The Ising model can be extended to include many-spin interactions, which can lead to drastic changes of its properties, and, in particular, to a dependence of the critical indices on the interaction energies (non-universality). The Ising model, as such, is a special case of the more general Heisenberg model, which for the three-dimensional case gives a phase transition with critical indices different from those of the Ising model. Another particular case of the Heisenberg model is the x–y model, with vortices as an essential part of it. The latter play a crucial role in the modern theory of superfluidity and superconductivity. Another pair of soluble models are the Gaussian and spherical models, with the critical indices in the spherical model different from those in other models. Nevertheless, all these models can be arranged by dividing them into a few universality classes. The university class to which a system belongs is specified by the following three factors:

(1) The spatial dimension d. In this respect, the lower critical dimensions, i.e., the minimum dimensions for the occurrence of phase transitions, are two for the Ising and three for the Heisenberg model. The upper critical dimension for these models is four. Indeed, for four and higher dimensions the phase transitions are described by mean field theory, with its own universality class.

(2) The dimension (number of components) D of the order parameter. This dimension is unity for the Ising model and three for the Heisenberg model, and so these two models lead to different critical behavior in three-dimensional systems.

(3) The finite or infinite radius of the interaction. In the latter case, each object interacts with all the others in the system, and one can neglect fluctuations (a situation similar to that in high-dimensional systems) which immediately leads to the mean field universality class.

All other factors, which define the short-range behavior (number of states at each site, type of the crystal lattice, etc.) of the system, do not influence the critical behavior.

Chapter 9

Phase Transitions in Reactive Systems

9.1 Multiple Solutions of the Law of Mass Action

Thus far, we have considered phase transitions in thermodynamic systems near their critical point. Another interesting problem is phase transitions in reactive systems as compared to non-reactive systems. Will the presence of a chemical reaction induced, say, by a very small amount of catalyst, stimulate or restrict the process of phase transition? In this chapter we will consider these problems, but we first have to consider the possibility of multiple solutions of the law of mass action. Two (or more) coexisting phases which have the same temperature and pressure are different in concentrations of different components and, therefore, in conducting chemical reactions. This means that the law of mass action has more than one solution for the concentration at given temperature and pressure.

Let us clarify the possibility of multiple solutions of the law of mass action. The chemical potential of the i-th component can be written in the following form

$$\mu_i = \mu_i^0\,(p, T) + \kappa_B T \ln\left(\gamma_i, x_i\right)\,, \tag{9.1}$$

where the activity γ_i determines the deviation from the ideal system, for which $\gamma_i = 1$. Using (9.1), one can rewrite the law of mass action as

$$x_1^{\nu_1}, x_2^{\nu_2}, ... x_n^{\nu_n} = K_{id.}\,(p, T)\ \gamma_1^{-\nu_1}, \gamma_2^{-\nu_2}, ... \gamma_n^{-\nu_n} \equiv K\,. \tag{9.2}$$

The chemical equilibrium constant for the ideal system K_{id} is determined by the function μ_i^0 in (9.1), i.e., by the properties of individual non-reactive components, whereas for non-ideal systems, K also depends on the interactions among the components. For ideal systems, all $\gamma_i = 1$, and Eq. (9.2) has a single set of solutions $x_1, x_2, ...x_n$. This was proved many years ago [72] for isothermal-isochoric systems. Another proof was given for isobaric systems under isothermal [73] and adiabatic [74] conditions. A complete analysis has recently been carried out [75].

It worth mentioning that phase transition is possible in ideal, although slightly artificial systems, such as that for which n units (atoms, molecules) of type A can reversibly form an aggregate (molecule, oligomer) of type A_n. Then, for large n (in fact, $n \to \infty$), the law of mass action has more than one solution, indicating the possibility of phase separation. It was subsequently noticed [76] that the latter effect is a special case of the phenomenon of the enthalpy–entropy compensation. This implies that for chemical reactions exhibiting a linear relationship between enthalpy and entropy, the magnitude of change in the Gibbs free energy is less than one might expect. Mathematically, in the Gibbs free energy equation ($\Delta G = \Delta H - T\Delta S$), the change in enthalpy (ΔH) and the change in entropy (ΔS) have opposite signs. Therefore, ΔG may change very little even if both enthalpy and entropy increase. However, the existence of the enthalpy–entropy compensation effect has been doubted by some researchers, as one can see from the article entitled "Enthalpy–entropy compensation: a phantom phenomenon" [77].

In general, a system has to be considerably non-ideal for the existence of multiple solutions of the law of mass action (9.2), i.e., the interaction energy must be on the order of the characteristic energy of an individual component. One possibility is for the interaction energy between components to be high. Consider a gas consisting of charged particles (plasma, electrolytes, molten salts, metal–ammonia solutions, solid state plasma). Another possibility occurs when the characteristic energies of single particles are small, as in the case of isomers. Both the isomerization [78] and the dissociation [79] reactions have been analyzed in detail.

Consider the ionization equilibrium of the chemical reaction of the form $A \rightleftarrows i + e$ (dissociation–recombination of neutral particles into positive and negative charges). Neglecting the complications associated with the infinite number of bound states, assume that $K_{id} \sim \exp\left(I_0/\kappa_B T\right)$, where I_0 is the ionization potential of a neutral particle. Assuming Debye screening of the electrostatic interactions, one readily finds [79], [80] that

$$K = K_{id} \exp\left[-\Phi\left(x, T\right)/\kappa_B T\right] \approx \exp\left\{\left[I_0 - \Phi\left(x, T\right)\right]/\kappa_B T\right\},$$

$$\Phi\left(x, T\right) \approx \left(8\pi x\right)^{1/2}\left(e^2/\kappa_B T\right)^{3/2} + Bx + \dots,$$

$$(9.3)$$

where x is the concentration of charges, and the function B, which determines the pair correlation between charges, has been tabulated [73].

At certain T and p, Eq. (9.3) has several solutions for x. This has a simple physical meaning. The function $\Phi\left(x, T\right)$ in (9.3) diminishes the ionization potential I_0 as a result of screening. Therefore, the phase with the larger degree of ionization has higher energy and higher entropy. Hence, these two phases can have equal chemical potentials and, therefore, can coexist. The chemical reaction proceeds in different ways in the two coexisting phases. Therefore, the appearance of multiple solutions of the law of mass action is a necessary condition for phase separation in reactive systems, where a chemical reaction proceeds in all phases.

9.2 Phase Separation in Reactive Binary Mixtures

Let us assume that two components A_1 and A_2 participate in a chemical reaction of the form $\nu_1 A_1 \longleftrightarrow \nu_2 A_2$. The law of mass action for this reaction is

$$A = \nu_1 \mu_1 + \nu_2 \mu_2 = 0,$$

$$(9.4)$$

where A is the affinity of reaction. If the system separates into two phases, their temperatures, pressures and chemical potentials are

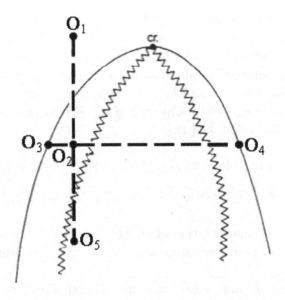

Fig. 9.1 The temperature–concentration phase diagram of a binary mixture. Quenching is performed from the original homogeneous state O_1 to the metastable state O_2 or to the non-stable state O_5. In the former case, phase separation to the final stable state, described by points O_3 and O_4, occurs through the nucleation process, whereas in the latter case it occurs through spinodal decomposition.

equal

$$\mu_1\left(p, T, x'\right) = \mu_2\left(p, T, x''\right), \qquad (9.5)$$

where x' and x'' denote the concentrations of one of the components in the two phases.

The phase diagram of a reactive binary mixture is shown in Fig. 9.1. The kinetics of phase separation from O_2 into two phase, described by points O_3 and O_4, proceeds in two clearly distinguishable stages. During the first stage, the system is located at point O_2, "waiting" for the appearance of a significant number of critical nuclei due to fluctuations (the duration of this stage is usually called "the lifetime of the metastable state"). At the end of this stage, the system starts to separate into two phases, and after some "completion" time, it reaches the two-phase state described by the points O_3 and O_4.

The existence of a chemical reaction(s) increases the stability of a metastable state through the appearance of additional constraint(s). Moreover, thermodynamic considerations give only necessary but not sufficient conditions for phase separation in reactive mixtures. Even when thermodynamics allows phase separation, it might be impossible from the kinetic aspect. The latter can be achieved in two different ways. The first way is, if the forward and backward reaction rates are different, with a preferred component corresponding to point O_3. Then, a chemical reaction leads to an additional driving force $A \neq 0$, shifting the system homogeneously from a quenched metastable state to the closest stable state O_3. The points O_3 and O_4 are separated by a singular line defining the stability boundary. Therefore, spatially homogeneous relaxation from O_2 to O_4 is impossible. Strictly speaking, for such reactive systems, the first stage of nucleation does not exist. Immediately after a quench from point O_1 to O_2, the system is shifted by chemistry to the closest state on the coexistence curve at the same temperature. The second possibility takes place when the forward and backward reactions rates are equal. If, in addition, the rate of reaction is much faster than that of phase separation, the chemical reaction, which tends to randomly mix the mixture, will keep the mixture homogeneous.

These physical arguments are supported by simple model calculations. Let us start with the simplest model of a strictly regular solution [81], where the chemical potentials have the following form [3]

$$
\begin{aligned}
\mu_1 &= \mu_1^0 + \kappa_B T \ln x + \omega \left(1 - x\right)^2 , \\
\mu_2 &= \mu_2^0 + \kappa_B T \ln \left(1 - x\right) + \omega x^2 .
\end{aligned}
\tag{9.6}
$$

The coexistence curve is defined by Eq. (9.5). Using (9.6) one gets for the concentrations of the two coexistence phases [3], $x \equiv x' = 1 - x''$, and the following symmetric coexistence curve,

$$
\frac{\omega}{\kappa_B T} = \frac{\ln x - \ln \left(1 - x\right)}{2x - 1} .
\tag{9.7}
$$

Using (9.4), the law of mass action can be written as

$$\frac{\omega}{\kappa_B T} = \frac{\ln x - \delta \ln (1 - x)}{\Lambda + \delta (1 - x)^2 - x^2}, \tag{9.8}$$

where

$$\delta = -\frac{\nu_1}{\nu_2}; \qquad \Lambda = \frac{\delta \mu_1^0 - \mu_2^0}{\omega}. \tag{9.9}$$

Phase separation in a reactive binary mixture will occur if and only if Eqs. (9.7) and (9.8) have a common solution. One can immediately see that Eqs. (9.7) and (9.8) coincide if both $\Lambda = 0$ ("symmetric" mixture) and $\delta = 1$ (isomerization reaction). If $\delta = 1$, $\Lambda \neq 0$, these equations have no common solution, and the strictly regular reactive binary mixture will not separate into two phases. In all other cases, one can equate the right-hand sides of Eqs. (9.7) and (9.8) to obtain

$$\frac{\left[(1 - x)^2 \ln (1 - x) - x^2 \ln x \right]}{\ln x - \ln (1 - x)} = \frac{\Lambda}{\delta - 1} \equiv q. \tag{9.10}$$

The concentration x ranges from zero to unity. Therefore, Eq. (9.10) has a solution only for $q < (2 \ln 2 - 1)/4 = 0.097$. The model of a strictly regular reactive binary mixture allows phase separation only for substances and chemical reactions satisfying this inequality. If, for example, $\Lambda = 0.09$ and $\delta = 2$, the criterion $q < 0.097$ is satisfied, two phases may coexist, and their concentrations are given by intersection of the coexistence curve (9.7) and the law of mass action (9.8). By contrast, for $\Lambda = 0.02$ and $\delta = 1$, this criterion is not satisfied, and hence there is no phase separation.

The non-compatibility of Eqs. (9.7) and (9.8) for some range of temperatures and pressures, i.e., the nonexistence of a common solution for the concentrations in the range from zero to unity, may appear in some models of an n-component system with $n - 1$ independent chemical reactions. In all other cases, the existence of r chemical reactions will decrease by r the dimension of the coexistence hypersurface.

Phase separation in binary mixtures has been described by models more complex than that of the strictly regular solution, such as the van Laar and Margules models [82].

The calculations performed above for a chemical reaction in a strictly regular binary mixture can easily be generalized to a three-component mixture of components $A + B \rightleftarrows C$, described by the activity coefficients

$$\ln \gamma_1 = \omega x_2 \left(1 - x_1\right); \; \ln \gamma_2 = \omega x_1 \left(1 - x_2\right); \; \ln \gamma_3 = -\omega x_1 x_2 . \quad (9.11)$$

If there are two independent chemical reactions, a system can only exist in one phase. However, when there is a single reaction, the situation is very different. A complete analysis has been given for a reversible bimolecular reaction that involves all three species [83].

9.3 Thermodynamic Analysis of Reactive Ternary Mixtures

Another way to find the global phase diagram for reactive ternary mixtures is by topological analysis of the intersection of the critical manifolds of the model considered and the specified chemical equilibrium surface. Let us start with the mean-field model described by Eq. (9.1) with activity coefficients given by (9.11) [84]. Phase equilibrium is obtained by demanding that all three chemical potentials be equal in coexistence phases. The phase equilibrium surface has the following form,

$$\frac{\omega}{\kappa_B T} = \frac{1}{x_1 - x_2} \ln \frac{x_1}{x_2}, \quad (9.12)$$

and the critical curve is given by the limit $x_1 - x_2 \to 0$,

$$x_{1,cr} = x_{2,cr} = \frac{\kappa_B T_{cr}}{\omega} . \quad (9.13)$$

The constraint $A = \mu_1 - \mu_2 - \mu_3 = 0$, imposed by the chemical reaction, takes the following form

$$x_3 = \exp \left(-\frac{\Delta G_0}{\kappa_B T}\right) x_1 x_2 \exp \left[\frac{\omega \left(x_1 + x_2 - x_1 x_2\right)}{\kappa_B T}\right], \quad (9.14)$$

where ΔG_0 is the deviation of the standard Gibbs free energy from a hypothetical ideal solution state, which can be written as $\Delta G_0 = \Delta H_0 - T\Delta S_0$, with ΔH_0 and ΔS_0 being the standard enthalpy and entropy for the given reaction. The intersection of this chemical equilibrium surface with the coexistence surface (9.12) defines a unique coexistence curve in the plane $(\kappa_B T/\omega,\ x_1)$ for fixed values of $\Delta G_0/\kappa_B T$. Numerical calculations for specific values of $\Delta G_0/\kappa_B T$ show [84] that $A - B$ repulsion results in phase separation as the temperature is lowered. Upon further lowering of the temperature, the formation of C becomes favorable, thereby reducing the unfavorable mixing, and the phases become miscible again. At low enough temperatures, the solution consists only of C and either A or B. Hence, the presence of a chemical reaction might result in the appearance of the lower critical consolute point which, together with an existing upper critical consolute point, results in a close-loop coexistence curve.

Another peculiarity of the phase diagrams in reactive ternary mixtures has been found [85] in the context of a slightly more complex model for $A + B \rightleftarrows C$ reaction, in which the chemical potentials of the components have the following form,

$$\mu_1 = \mu_1^0 + \kappa_B T \ln x_1 + bx_3 + cx_2 - W,$$

$$\mu_2 = \mu_2^0 + \kappa_B T \ln x_2 + ax_3 + cx_1 - W, \qquad (9.15)$$

$$\mu_3 = \mu_3^0 + \kappa_B T \ln x_3 + ax_2 + bx_1 - W,$$

where $W = ax_2x_3 + bx_1x_3 + cx_1x_2$. The chemical equilibrium constraint $A = 0$ for Eqs. (9.15) can be expressed as follows,

$$\kappa_B T \ln \frac{x_3}{x_1 x_2} = a\,(x_3 - x_2) + b\,(x_3 - x_1)$$

$$+ c\,(x_1 + x_2) - W - \Delta H_0 + T\Delta S_0. \qquad (9.16)$$

The intersection of this surface and the critical manifolds following from (9.15) gives the phase diagram of a reactive ternary mixture. The results depend on whether the chemical process is enthalpically or entropically favored. Detailed analysis shows [85] that there are six different types of phase diagrams, depending on the combinations of sign of the parameters a, b and c which include the triple and quadruple points, closed-loop phase coexistance curves, azeotrope-like points, etc.

As an example of homogeneous nucleation in a chemically reactive system, let us consider [86] a simple bistable chemical reaction (the second Schlogl trimolecular model),

$$A + X \underset{k_2}{\overset{k_1}{\rightleftarrows}} 3X; \qquad X \underset{k_4}{\overset{k_3}{\rightleftarrows}} B \qquad (9.17)$$

with the concentrations x_A and x_B of species A and B being held constant. The kinetic equation for the concentration x of the component C is

$$\frac{dx}{dt} = -k_1 x^3 + k_2 x_A x^2 - k_3 x + k_4 x_B = -\frac{dV(x)}{dx}. \qquad (9.18)$$

Equation (9.18) has three solutions. We assume that the values of the coefficients k_i provide two stable solutions x_1 and x_3, and one unstable solution x_2, such that $x_1 < x_2 < x_3$. Moreover, it is suggested that $V(x_1) < V(x_3)$, i.e., the state x_3 is stable while the state x_1 is metastable. If the reaction (9.17), supplemented by diffusion, occurs in a closed one-dimensional vessel, Eq. (9.18) will have the form of the reaction-diffusion equation

$$\frac{\partial x(\rho, t)}{\partial t} = -k_1 x^3 + k_2 x_A x^2 - k_3 x + k_4 x_B + D\frac{\partial^2 x}{\partial \rho^2}, \qquad (9.19)$$

where ρ is the spatial coordinate and D is the diffusion coefficient. Introducing the dimensionless variables $u_i = x_i/x_1$; $\tau = t k_2 x_1^2$; $D = \mathcal{D}/k_1 x_1^2$, and moving the origin to the metastable state x_1, $\zeta = (x - x_1)/x_1$, one gets

$$\frac{\partial \zeta}{\partial \tau} = D\frac{\partial^2 \zeta}{\partial \rho^2} - b\zeta + a\zeta^2 - \zeta^3, \qquad (9.20)$$

with $b = (u_2 - 1)(u_3 - 1)$ and $a = u_2 + u_3 - 2$. If the system is quenched to the metastable state $\zeta = 0$, its transition to the stable state $\zeta = \zeta_3$ can be understood as the spontaneous creation of two solutions of Eq. (9.20) ("kink"–"antikink")

$$\zeta\left(\rho,t\right) = \zeta_3 \left\{1 + \exp\left[\pm\delta\left(\rho - \rho_0 \pm vt\right)\right]\right\}^{-1}, \qquad (9.21)$$

where

$$\zeta_3 = \frac{1}{2}\left[a + \left(a^2 - 4b\right)^{1/2}\right]; \quad \delta^2 = \frac{\zeta_3^2}{2D}; \quad v = -\frac{3b - a\zeta_3}{2\delta}. \qquad (9.22)$$

The kink–antikink pair separates with time in opposite directions. When separated by a certain critical length κ, equal to $-\delta^{-1}\ln\left[\left(a - 3\sqrt{b/2}\right)\left(a + 3\sqrt{b/2}\right)^{-1}\right]$, they create a stationary profile, called a nucleation nucleus $\zeta\left(\rho\right)$. When the kink and the antikink are separated by a distance less than κ, they annihilate, and the concentration profile returns to the homogeneous metastable state. When the separation becomes greater than κ, the kink–antikink pair separates, and the system evolves toward the homogeneous stable state. Equation (9.21) becomes simplified in the small-amplitude-nucleus limit, when $\kappa \to 0$, and the nucleation nucleus then becomes

$$\zeta\left(\rho\right) = \frac{3b}{a}\left[1 + \cosh\left(\delta\rho\right)\right]^{-1} = \frac{3b}{2a}\ \text{sech}^2\left(\frac{\delta\rho}{2}\right). \qquad (9.23)$$

Thus far, we have discussed a deterministic system. The influence of homogeneous external white noise and internal chemical fluctuations has been considered [87] in the framework of the extended Kramers approach and the multivariate master equation, respectively. The nucleation due to the internal fluctuations is limited to a very narrow region near the transition, and it is difficult to observe this experimentally. The external noise may be easily greater than the internal noise and, therefore, more important. In addition to the homogeneous nucleation initiated by the fluctuations (considered above), the important problem is non-homogeneous nucleation initiated, say, by pieces of dust present in solutions.

9.4 Kinetics of Phase Separation

Let us first cosider non-reactive systems. All thermodynamic states within the coexistence curve shown by the solid line in Fig. 9.1 belong to two-phase states. Therefore, after a change in temperature (or pressure) from point O_1 to O_2, phase separation will occur and the system will consist of two phases described by the thermodynamic condition (9.5) and shown by points O_3 and O_4 in the figure. All states located within the coexistence curve and close to it are the so-called metastable states. Although the energy of these states is larger than that of the corresponding two-phase system, one needs a finite "push" to pass from the local minimum in energy to the deeper global minimum. Such a "push" is provided by the thermal fluctuations in density for a one-component system and in concentration for many-component mixtures. Therefore, these fluctuations are of crucial importance for the analysis of the decay of a metastable state.

The change of the thermodynamic potential associated with the appearance of the nucleus of the new phase (assumed, for simplicity, to be spherical with radius r) is equal to

$$\Delta G\left(r\right) = -\Delta\mu\frac{4}{3}\pi r^3 + 4\pi\sigma r^2 , \qquad (9.24)$$

where $\Delta\mu$ is the energy gain in a stable state compared to a metastable state, and σ is the surface tension. The first (negative) term in (9.24) encourages phase separation while the second (positive) term prevents it. It is clear from (9.24) that for small nuclei, $\Delta G > 0$. That is, small nuclei tend to shrink because of their high surface-to-volume ratio. Only nuclei larger than the critical size r_c are energetically favorable. From $d\left(\Delta G\right)/dr = 0$, one finds

$$r_c = \frac{2\sigma}{\Delta\mu} . \qquad (9.25)$$

The traditional phenomenological approach to the decay of a metastable state of a pure substance is based on the distribution function $W\left(r, t\right)$ of nuclei of size r at time t. The continuity equation has the form

$$\frac{\partial W}{\partial t} = \frac{\partial}{\partial r}\left[FW + D\frac{\partial W}{\partial r}\right] \equiv -\frac{\partial J}{\partial t}, \qquad (9.26)$$

where the flux $J(r,t)$ of nuclei along the size axis is determined by two unknown functions F and W. One can reduce [87] the equation(s) of the critical dynamics to the Langevin equation for the radius of nuclei with a known random force, which can be transformed to the Fokker–Planck equation (9.26) with functions F and D of the form

$$F(r) = D_0\left(\frac{1}{r} - \frac{1}{r_C}\right); \qquad D(r) = \frac{D_0\kappa_B T}{8\pi\sigma r^2}, \qquad (9.27)$$

where D_0 is the diffusion coefficient far from the critical point.

The nucleation process in a binary mixture can be described in analogous fashion [88]. The formation energy of a nucleus containing n_1 atoms of component 1 and n_2 atoms of component 2 is

$$\Delta G = (\mu_{1,n} - \mu_1)\,n_1 + (\mu_{2,n} - \mu_2)\,n_2 + 4\pi\sigma r^2, \qquad (9.28)$$

where μ_1 and μ_2 are the chemical potentials of the components in the homogeneous phase, and $\mu_{1,n}$ and $\mu_{2,n}$ are the corresponding quantities in the nuclei. The size r of a nucleus is related to its structure by $4\pi r^3/3 = v_1 n_1 + v_2 n_2$, where v_1 and v_2 are the volumes per atom.

In contrast to a pure substance, the critical nucleus is defined not only by its size r, but also by the concentration $x_1 = n_1/(n_1 + n_2)$. Therefore, the height of the potential barrier is defined by $\partial\Delta G/\partial r = \partial\Delta G/\partial x = 0$, which gives the following relation between r_C and x_C of the critical nucleus,

$$r_C = 2\sigma\frac{x_C v_1 + (1 - x_C)\,v_2}{x_C\Delta\mu_1 + (1 - x_C)\,\Delta\mu_2}, \qquad (9.29)$$

where $\Delta\mu_i = \mu_{i,n} - \mu_i$.

Let us now consider reactive systems. The concentration x changes as a result of a chemical reaction, which shifts the initial quenched state to the closest equilibrium state on the coexistence curve. Each intermediate state corresponds to a different radius of critical nucleus $r_C(t)$, increasing toward the coexistence curve, where

$r_C \to \infty$. Hence, one has to replace r_C by $r_C(t)$ everywhere. The latter function can be found from Eq. (9.29) under the assumption of a quasi-static chemical shift. This shift is caused by a chemical reaction, and can be described in the linear approximation by

$$\frac{dx}{dt} = -\frac{x - x'}{\tau}, \tag{9.30}$$

where τ^{-1} is the rate of the chemical reaction.

Inserting the solution of Eq. (9.30), $x - x' = (x_0 - x') \exp(-t/\tau)$, into Eq. (9.29), one obtains

$$r_C(t) = \frac{2\sigma \left[x'' v_1 + (1 - x'') v_2 \right]}{\kappa_B T \left[x''/x' - (1 - x'')/(1 - x') \right] (x' - x'')}$$

$$\equiv r_0 \exp(t/\tau), \tag{9.31}$$

where we used the coexistence condition (9.5), and retained only the leading logarithmic part of the chemical potentials.

The next step is to obtain the Fokker–Planck equation for the distribution function $W(r, t, x)$ for nuclei of size r and composition x at time t. We make the approximation that all nuclei which are important for phase separation have the same composition x''. Therefore, we may omit the argument x in $W(r, t, x)$. In other words, one assumes that a chemical reaction brings the path leading to the saddle point closer to that of $x = x''$.

Substituting (9.27) and (9.31) into the Fokker–Planck equation (9.26), one obtains

$$\frac{\partial W}{\partial t} = \frac{\partial}{\partial r} \left\{ D_0 \left[\frac{1}{r} - \frac{1}{r_0} \exp\left(\frac{t}{\tau}\right) \right] \right\} W + \frac{D_0 \kappa_B T}{8\pi\sigma r^2} \frac{\partial W}{\partial r}. \tag{9.32}$$

To find the approximate solution of this complicated equation, we first consider a simplified version, which allows an exact solution. One replaces functions $F(r)$ and $D(r)$ in (9.27) by simpler functions which incorporate the main property of (9.27). According to the physical picture, $F(r)$ is positive for $r < r_c$ and negative for

$r > r_c$. Therefore, we approximate $F(r)$ by

$$F(r) = B(r_c - r) .$$ (9.33)

After choosing this form of $F(r)$, we are no longer free to choose the second function $D(r)$. These two functions are not independent because there is no flux in equilibrium, which, according to (9.26), leads to $W_{eq} = \exp\left[-\int dr F(r)/D(r)\right]$. On the other hand, by definition, $W = \exp(-\Delta G/k_B T)$, and, therefore,

$$\int \frac{F(r)}{D(r)} dr = \frac{\Delta G}{k_B T} ,$$ (9.34)

where $\Delta G(r)$ and r_c are given by (9.24) and (9.25). A simple calculation yields

$$D(r) = \frac{B k_B T r_c}{8 \pi \sigma r} \approx \frac{B k_B T}{8 \pi \sigma} .$$ (9.35)

We keep only the largest term in $r - r_c$ in the last equality in (9.35). Using (9.33), (9.31) and (9.35), one obtains the simplified Fokker–Planck equation,

$$\frac{\partial W}{\partial t} = \frac{\partial}{\partial r} \left[B(r_0 \exp(t/\tau) - r) W \right] + D_0 \frac{\partial^2 W}{\partial r^2} ,$$ (9.36)

where $D_0 = B k_B T/8 \pi \sigma$. Equation (9.36) can be solved exactly by introducing the characteristic function

$$W(K, t) = \int W(r, t) \exp(iKr) dr .$$ (9.37)

The original Fokker–Planck equation (9.26)–(9.27) is thus transformed into a first-order linear partial differential equation for $W(K, t)$,

$$\frac{\partial W}{\partial t} - BK \frac{\partial W}{\partial K} = -\left[D_0 K^2 + iKBr_0 \exp\left(\frac{t}{\tau}\right) \right] W .$$ (9.38)

To solve this equation, one uses the method of characteristics, with initial condition

$$W(r, t = 0) = \delta(r - \xi) ,$$ (9.39)

where ξ is on the order of a single molecule (there are no nuclei immediately after a quench). Solving Eq. (9.38) and performing the inverse Fourier transformation, one obtains the Gaussian distribution function

$$W(r,t) = \frac{1}{\sqrt{4\pi \overline{\Delta r^2}}} \exp\left[-\frac{(r - \bar{r})^2}{2\overline{\Delta r^2}} \right], \tag{9.40}$$

with mean value $\bar{r}$ and variance $\overline{\Delta r^2}$ given by

$$\bar{r} = \xi \exp(Bt) - \frac{Br_0}{\tau^{-1} - B}\left[\exp\left(\frac{t}{\tau}\right) - \exp(Bt) \right],$$

$$\overline{\Delta r^2} = \frac{D_0}{2B}\left[\exp(2Bt) - 1 \right]. \tag{9.41}$$

Two characteristic times, B^{-1} and τ, are associated with transient processes and chemical reactions, respectively. However, the solution (9.41) diverges with time, so that Eq. (9.36) does not have a steady-state solution. Hence, instead of the exact solution, one has to use the approximate solution.

One can find [89] the appropriate solution of Eq. (9.36), but we prefer to find the solution of the original equations (9.26)–(9.27), assuming that B^{-1} is very small [90]. This assumption permits us to neglect transient processes, and, for $B^{-1} < \tau$, restrict attention to those solutions of (9.36) which supply the quasi-static flux $J_{qss}(t)$ independent on the size of the nuclei. We consider the quasi-steady-state regime which is established after the transients disappear. The quasi-steady-state solution $W_{qss}[r, r_c(t)]$ does not have an explicit time dependence, so that $\partial W_{qss}/\partial t = 0$, and the Fokker–Planck equation (9.36) can be rewritten as

$$D_0\left[\frac{1}{r} - \frac{1}{r_0}\exp\left(\frac{t}{\tau}\right) \right]W_{qss} + \frac{D_0\kappa_B T}{8\pi\sigma r^2}\frac{\partial W_{qss}}{\partial r} \equiv J_{qss}\left(\frac{t}{\tau}\right). \tag{9.42}$$

The quasi-steady-state flux J_{qss} reduces to the steady-state flux J_{ss} when the chemical reaction is absent,

$$J_{qss}\left(\frac{t}{\tau} = 0\right) = J_{ss}. \tag{9.43}$$

There is no stationary state for the reactive system considered here. Therefore, one is forced to give a new definition to the lifetime of a metastable state in reactive systems. The simplest generalization is the time required to produce one critical nucleus

$$\int\limits_{0}^{T_{ch}} dt\, J_{qss}\,(t) = 1\,.$$

(9.44)

For the time-independent case, J_{qss} is replaced by J_{ss}, according to (9.43), and T_{ch} is replaced by the lifetime of the metastable state in the non-reactive system, $T_0 = J_{ss}^{-1}$.

Let us now turn to the solution of Eq. (9.42). The boundary conditions for this equation are determined by the requirements that the distribution of nuclei of minimal size ξ will be the equilibrium distribution, and the total number of nuclei in the system is bounded,

$$W_{qss}\,(\xi) = W_{eq}\,(\xi)\,; \qquad W_{qss}\,(r \to \infty) = 0\,,$$

(9.45)

where W_{eq} corresponds to zero flux.

The solution of Eq. (9.42) which satisfies the boundary conditions (9.45) has the form [87],

$$W_{qss} = W_{eq}\,(r) \left[1 - J_{qss} \int\limits_{\xi}^{r} \frac{dz}{W_{eq}\,(z)\, D\,(z)} \right]\,,$$

(9.46)

where

$$J_{qss}^{-1} = \int\limits_{\xi}^{\infty} \frac{dz}{W_{eq}\,(z)\, D\,(z)}\,.$$

(9.47)

The function $W_{eq}^{-1}\,(r)$ has a sharp maximum at r_C, which reflects the existence of a barrier to nucleation. Therefore, the integral (9.47) can be evaluated by the method of steepest descents,

$$J_{qss} = \sqrt{\frac{\kappa_B T}{4\pi^2\sigma} \frac{D_0}{2r_0^2}} \exp\left\{-\frac{2t}{\tau} - \frac{\Delta G\left(r_0\right)}{\kappa_B T} \exp\left(\frac{2t}{\tau}\right)\right\}, \qquad (9.48)$$

where $\Delta G\left(r_0\right) = 4\pi\sigma r_0^3/3$ is the minimal work needed to produce the critical nucleus in the initial state immediately after a quench.

Upon inserting J_{ss}, defined by (9.43), into (9.48), one obtains

$$J_{qss} = J_{ss} \exp\left\{-\frac{2t}{\tau} - u\left[\exp\left(\frac{2t}{\tau}\right) - 1\right]\right\}, \qquad (9.49)$$

where $u = \Delta G\left(r_0\right)/\kappa_B T$.

Finally, inserting (9.49) and $T_0 = J_{ss}^{-1}$ into (9.44) gives the equation for the lifetime T_{ch} of a metastable state in a reactive system as a function of the lifetime in the non-reactive system T_0 (the latter depends on the volume under observation), the extent of quench u, and the rate τ^{-1} of the chemical reaction,

$$1 = \frac{\tau u \exp\left(u\right)}{2T_0} \int\limits_{u}^{u\exp(2T_{ch}/\tau)} \frac{\exp\left(-z\right)}{z^2} dz. \qquad (9.50)$$

Equation (9.50) for T_{ch} has been solved numerically. This equation has no solution if τ is very small, i.e., the chemical reaction is very fast. Then, although the thermodynamics allows phase separation in a reactive system, such a separation is impossible from the kinetic point of view. The system is dragged by the chemical reaction to a homogeneous equilibrium state on the coexistence curve before the nuclei of the new phase appear. The minimal τ which allows phase separation for different quenches (with characteristic u and T_0) is given approximately by the following formula,

$$\left(\frac{\tau}{T_0}\right)_{min} \approx 3.03 + 2.08u. \qquad (9.51)$$

The new phenomenon appears when the vortices are moving in a relatively thick (thicker than the magnetic penetration depth) type II superconducting sample under the influence of an external magnetic field. This field is able to penetrate into the sample and take part in

time-dependent processes. Its behavior is described by the equation
for vortex diffusion

$$\frac{\partial B}{\partial t} = \frac{c^2}{4\pi} \frac{\partial}{\partial x} \left[R(\Psi) \frac{\partial B}{\partial x} \right], \qquad (9.52)$$

where the effective diffusion coefficient $R(\Psi)$ depends on the order
parameter Ψ of the vortex system, describing the difference between
the diffusion of the magnetic induction in the ordered and disordered phases. In contrast to the case of a superconducting film [91]
(where one can consider a constant magnetic field along the sample), in the present case of a superconducting slab, one has to solve
the Ginzburg–Landau equation together with the diffusion equation (9.52). The simultaneous solution of these two equations [92]
reveals the new phenomenon of oscillating magnetic induction ("flux
waves"). A similar phenomenon (where the diffusion equation with
monotonically decaying solutions, being coupled with the second
equation, which has a wave-type solutions) occurs in the description
of convection by the Navier–Stokes and heat conductivity equations.
Then, for a critical value of the temperature gradient, one obtains the
frequency and wave vector of an oscillating convective solution. In
our case, the role of the temperature gradient is played by the order
parameter Ψ. Partially, these theoretical calculations agree with the
magneto-optical measurements on BSCCO crystal [93].

9.5   Change of Critical Parameters Due to a Chemical
Reaction

Chemical reactions near the critical points are routinely studied near
the consolute points of a binary mixture which plays the role of a
solvent. The presence of a reactant, as well as impurities, pressure
or an electric field changes the critical parameters of a binary mixture. There is only limited knowledge of the relationship between the
shifts of the critical parameters when the perturbation is applied. For
example, it is known [3] that an impurity will cause a critical temperature shift equal to that of the critical composition only when
the solubilities in the two components are roughly equal. However, it

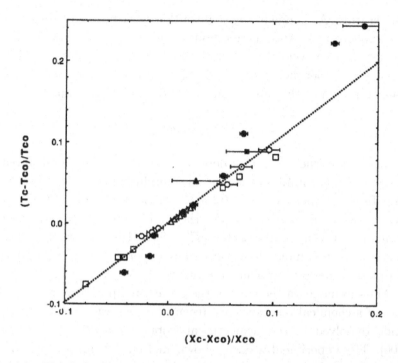

Fig. 9.2   The shift of critical points $T_{C,0}$ and $x_{C,0}$ of different systems to new values T_C and x_C. The different symbols refer to five different systems (for details, see [96]).

was found that two different impurities (water [94] and acetone [95]), which have quite different solubilities in the two components of the methanol–cyclohexane mixture, induce equal relative shifts in critical temperature and composition, $\Delta T_C/T_{C,0} = \Delta x_C/x_{C,0}$. The same relation has been observed [96] in a wide range of systems (Fig. 9.2). These data have been obtained by isobaric measurements upon the addition of impurities.

There are different experimental ways of finding the critical parameters T_C and p_C of a system. The simplest, but less precise method is based on visual detection of critical opalescence to determine the critical temperature. For ionic conductors, the critical temperature is identified by the sharp change in electrical conductivity [97]. Alternatively, one can use the dilatometric method of measuring the height of the liquid in the capillary side-arm as a function of temperature. A change in the slope of this curve occurs at the critical temperature [98]. Another method is the acoustic technique, using the fact that the velocity of sound reaches a minimum at the critical point [99]. Experiments show that the critical temperature T_C is a linear function of the initial concentration x of a reactant,

$$T_C = T_C^0 + ax. \tag{9.53}$$

The sign of the constant a depends on whether the chemical reaction enhances or inhibits the mutual solubility of the original solvent pair. In the former case, $a < 0$ for UCST, and $a > 0$ for LCST, and vice versa for the latter case. For the equilibrium mixture of benzyl bromide in triethylamine–water $\left(T_C^0 = 291.24 \text{ K}\right)$, it was found [97] that $a = 36.9$ K/mole. There are many experimental results related to different many-component mixtures [99].

Measurements of the shift of the critical temperature in the presence of a chemical reaction have recently been performed for indium oxide dissolved in the near-critical isobutyric acid–water mixture [100]. The experimental data, shown in Fig. 9.2, are described by Eq. (9.53) with $a < 0$, i.e., dissolving In_2O_3 in isobutyric acid–water lowers the critical solution temperature.

9.6 Modification of the Critical Indices

In this section, we describe the changes in critical phenomena arising from the presence of chemical reactions. The singularities of the thermodynamic and kinetic quantities near the critical points are determined by the critical indices. The calculations of these indices assumes the constancy of some intensive variables, say, pressure p or chemical potentials μ, whereas in practice it is impossible, for example, to ensure the constancy of μ (one has to vary the concentrations during the course of the experiment). Similarly, p remains practically unchanged near the liquid–liquid critical point since the experiment is carried out in the presence of saturated vapor. On the other hand, it is quite difficult to ensure the constancy of p near the liquid–gas critical point where a fluid is highly compressible.

In order to compare theoretical calculations for an ideal one-component system with experimental data obtained for "real" systems, Fisher [101] developed the theory of renormalization of critical indices. The main idea of renormalization can be explained by the following simple argument. For a pure substance, mechanical stability is determined by $(\partial p / \partial v)_T < 0$. The condition for diffusion stability in a binary mixture has the form $(\partial \mu / \partial x)_{T,p} > 0$, which, by a simple thermodynamic transformation, can be rewritten as $(\partial p / \partial v)_{T,\mu} < 0$. Therefore, at constant chemical potential μ, the critical behavior of a binary mixture, which is defined by the stability conditions, will be the same as that of a pure substance. Multicomponent systems behave in an analogous manner, namely, the stability condition for a n-component mixture is determined by the condition $(\partial p / \partial v)_{T,\mu_1 \dots \mu_{n-1}} < 0$, where $n - 1$ chemical potentials are held constant. However, the critical parameters now depend on the variables μ_i, and in order to obtain experimentally observable quantities which correspond to constant concentration x, one has to pass from $T(\mu)$ to $T(x)$. As a result, when one goes from a pure substance to a binary mixture, the critical indices above and below a critical point are multiplied by $\pm (1 - \alpha)^{-1}$ [101], where α is the critical index of the specific heat at constant volume. The minus sign refers to the specific heat at constant volume and the plus sign refers to all other critical indices. We have discussed [102] the renormalization of the critical indices associated with the presence of chemical reactions. As an example, let us again consider a binary fluid mixture $A_1 - A_2$ with the single reaction $\nu_1 A_1 + \nu_2 A_2 = 0$ (say, the isomerization reaction). This equation means that the law of mass action is satisfied, i.e., one gets for the extent of reaction $A = \nu_1 \mu_1 + \nu_2 \mu_2 = 0$, which, in turn, reduces by one the number of thermodynamic degrees of freedom in a binary mixture. Therefore, the reactive binary system has an isolated critical point, and the critical indices of this system are the same as those of a pure fluid,

$$\left(\frac{\partial v}{\partial p}\right)_{T,A=0} \sim C_{p,A=0} \sim \left(\frac{T-T_C}{T_C}\right)^{-\gamma},$$

$$\left(\frac{\partial v}{\partial p}\right)_{s,A=0} \sim C_{v,A=0} \sim \left(\frac{T-T_C}{T_C}\right)^{-\alpha}, \tag{9.54}$$

where $v = 1/\rho$ is the specific volume.

It is interesting to compare (9.54) with the case of a frozen chemical reaction (no catalyst added). The system considered is then a binary mixture with a liquid–gas critical line. The renormalized critical indices of such systems are well known [101]

$$C_{p,\xi} \sim \left(\frac{\partial v}{\partial p}\right)_{T,\xi} \sim \left(\frac{T-T_C}{T_C}\right)^{-\alpha/(1-\alpha)},$$

$$C_{v,\xi} \sim \left(\frac{\partial v}{\partial p}\right)_{s,\xi} \sim \left(\frac{T-T_C}{T_C}\right)^{\alpha/(1-\alpha)}. \tag{9.55}$$

The correspondence between (9.54) and (9.55) becomes obvious from the thermodynamic relation [3]

$$C_{p,A=0} = C_{p,\xi} - h^2 \left(\partial\xi/\partial A\right)_{T,p}, \tag{9.56}$$

where h is the heat of reaction. In fact, the asymptotic behavior of the left-hand side of Eq. (9.56) on approaching a critical point is determined by the second term on the right-hand side rather than by the first term, which has a weaker singularity.

The occurrence of a chemical reaction in the system under consideration modifies the critical indices of the observable specific heats at constant volume and constant pressure (or inverse velocity of sound) compared to a system with a frozen chemical reaction. They are changed from $\alpha/(1-\alpha)$ and $-\alpha/(1-\alpha)$ to $-\alpha$ and $-\gamma$, respectively. For example, the specific heat at constant volume has a weak singularity at the critical point when a chemical reaction occurs, rather than a finite, cusped behavior in the absence of a chemical reaction.

The singularities become weaker for many-component mixtures. A common situation is that the solutes undergo various chemical transformations while the solvent does not. As an example, consider the critical system containing a reactive binary mixture dissolved in some solvent. Due to the existence of a neutral third component, the system has a line of critical points rather than an isolated critical point, as is the case for a reactive binary mixture.

Let us consider the frozen chemical reaction. The singularities of the thermodynamic quantities in a ternary mixture at constant chemical potential μ_0 of the solvent are similar to those of a binary mixture. The liquid–liquid critical points depend only slightly on the pressure, so that the parameter $R\rho_C (dT_C/dp)$ (ρ_C is the critical density) is very small. This parameter determines the region of renormalization [103]. Therefore, renormalization is absent, as in the vicinity of the λ line in helium, and

$$C_{p,\xi,\mu_0} \sim C_{v,\xi,\mu_0} \sim \phi(\mu_0)^{-\alpha}; \quad \phi(\mu_0) \equiv \frac{T - T_C(\mu_0)}{T_C(\mu_0)}. \qquad (9.57)$$

However, measurements are taken at constant number of solvent particles rather than at $\mu_0 = $ const. According to renormalization, near the liquid–liquid critical points, one obtains

$$C_{p,\xi,N_0} \sim C_{v,\xi,N_0} \sim \left(\frac{T - T_C}{T_C}\right)^{\alpha/(1-\alpha)}. \qquad (9.58)$$

Unlike the liquid–liquid critical point, there are two renormalizations in the vicinity of liquid–gas critical point: the first occurs when passing from the binary to the ternary mixture, and the second renormalization takes place when passing from $\mu_0 = $ const to $N_0 = $ const. In the region of renormalization of the ternary mixture, one gets

$$C_{p,\xi,\mu_0} \sim \phi^{-\alpha/(1-\alpha)}; \quad C_{v,\xi,\mu_0} \sim \phi^{\alpha/(1-\alpha)}, \qquad (9.59)$$

and for $N_0 = $ const,

$$C_{p,\xi,N_0} \sim \phi^{\alpha/(1-\alpha)}; \quad C_{v,\xi,N_0} \sim \text{const}. \qquad (9.60)$$

Comparing (9.58) and (9.60), one concludes that without a chemical reaction, cusp-like behavior exists for both specific heats near the liquid–liquid critical points, but only for the specific heat at constant pressure near the liquid–gas critical points.

When a chemical reaction takes place, the singularities near the critical points can be found from the thermodynamic relations, analogous to (9.56),

$$
\begin{aligned}
C_{p,N_0,A=0} &= C_{p,N_0,\xi} - T\,(\partial\xi/\partial A)_{T,p,N_0}\,(\partial A/\partial T)^2_{p,N_0,\xi}\,, \\
C_{v,N_0,A=0} &= C_{v,N_0,\xi} - T\,(\partial\xi/\partial A)_{v,T,N_0}\,(\partial A/\partial T)^2_{v,N_0,\xi}\,.
\end{aligned}
\tag{9.61}
$$

For both types of critical points, the singularities of thermodynamic quantities in a system undergoing a chemical reaction ($A = 0$) are determined by the second term on the right-hand sides of Eqs. (9.61). The factor $\partial A/\partial T$ in the latter terms remains finite at the critical point, whereas the second factor has the following asymptotic behavior,

$$
(\partial\xi/\partial A)_{T,p,\mu_0} \sim \phi^{-\gamma}; \qquad (\partial\xi/\partial A)_{v,T,\mu_0} \sim \phi^{-\alpha}
\tag{9.62}
$$

and, after renormalization,

$$
\begin{aligned}
(\partial\xi/\partial A)_{T,p,N_0} &\sim \left(\frac{T-T_C}{T_C}\right)^{-\alpha/(1-\alpha)} \\
(\partial\xi/\partial A)_{v,T,N_0} &\sim \left(\frac{T-T_C}{T_C}\right)^{\alpha/(1-\alpha)}
\end{aligned}
\tag{9.63}
$$

Therefore, for a system undergoing a chemical reaction, one obtains from (9.61) and (9.63)

$$
\begin{aligned}
C_{p,N_0,A=0} &\sim \left(\frac{\partial v}{\partial p}\right)_{T,N_0,A=0} \sim \phi^{-\alpha/(1-\alpha)}\,, \\
C_{v,N_0,A=0} &\sim \left(\frac{\partial v}{\partial p}\right)_{v,N_0,A=0} \sim \phi^{\alpha/(1-\alpha)}\,.
\end{aligned}
\tag{9.64}
$$

Comparing (9.58) and (9.60) with (9.64) yields that the existence of a chemical reaction leads to a magnification of the singularities of the specific heats at constant pressure for both types of critical points, from cusp-like behavior $\alpha/(1-\alpha)$ to a weak singularity

$-\alpha/(1-\alpha)$. On the other hand, the specific heat at constant volume changes its asymptotic behavior from constant to cusp-like only near the liquid–gas critical point. In principle, one can detect such a magnification experimentally. Experiments are slightly easier to perform near the liquid–liquid critical points, because these points usually occur at atmospheric pressure and room temperature.

Consider now the singularities of the dielectric constant ε and electrical conductivity σ at the critical point of reactive systems. These quantities can be obtained by choosing poorly conducting liquid mixtures for dielectric measurements and strongly conducting mixtures for conductivity measurements. One can also measure ε and σ simultaneously placing the fluid between two parallel plates or coaxial cylindrical electrodes with an alternative current bridge. The fluid then acts in the bridge circuit as a frequency-dependent impedance with balanced capacity and resistive components [104]. Near the critical points, large fluctuations affect all thermodynamic and kinetic properties including those which are characterized by σ and ε. There are hundreds of experimental papers, of which we will describe only a few. Of special importance are chemical reactions in ecological clean near-critical water. At high temperature and pressure, water becomes self-ionized with a high conductivity, resembling molten salts [105]. The special case is the dissociation of a weak acid in water near an acid–water liquid–liquid critical point,

$$HA+H_2O \rightleftarrows A^-+H_3O^+ . \qquad (9.65)$$

Measurements of the conductivity as a function of temperature have been performed near the liquid–liquid critical points for the systems isobutyric acid–water [106], [107] and phenol–water [107]. Different explanations have been proposed for the anomaly in the conductivity near the critical point [108]: 1) The anomaly is due to viscous drag of the acid anion [106] and, hence, it is proportional to the inverse of the viscosity η, $\sigma \sim \eta^{-1} = \tau^{\nu z_\eta} = \tau^{0.031}$, where $\tau = (T-T_C)/T_C$, and $\nu = 0.063$, $z_\eta = 0.05$ are the critical indices for the correlation length and viscosity, respectively; 2) The anomaly of σ is related to the anomaly in the proton-transfer rate [109], which, in turn [110], has the temperature dependence of the nearest-neighbor correlation function, $\sigma \sim \tau^{1-\alpha} = \tau^{0.89}$, where $\alpha = 0.11$ is the critical index for the heat capacity at constant pressure and composition; 3) A percolation model for conductivity leads [107] to an anomaly of the form $\sigma \sim \tau^{2\beta} = \tau^{0.65}$, where $\beta = 0.325$ is the critical index describing the coexistence curve; 4) The anomaly of σ is related to that of the extent of the acid dissociation [111], which leads [112] to $\sigma \sim \tau^{1-\alpha} = \tau^{0.89}$.

A detailed analysis of the experimental data obtained in [106], [107] has been carried out [113], taking account of the background contribution due to the normal temperature dependence of the dissociation constant and due to the confluent critical singularity. The leading critical anomaly of the conductance was found to be characterized by a critical index $1 - \alpha$, consistent with its connection with an anomaly in the extent of the dissociation reaction and also for an anomaly in the proton-transfer rate.

The static dielectric constant for the dimerization reaction $2NO_2 \rightleftharpoons N_2O_4$ has been measured [114] in a solvent mixture of perfluoromethylcyclohexane–carbon tetrachloride near the liquid–iquid critical point of the solvent. The appropriate choice of reaction (no dipole moment for the N_2O_4 molecule, and a small moment for NO_2) and of the solvent (the components of low polarity make the "background" dipole moment very small) assure the high precision of measurements. The dielectric constant was found to have a $1 - \alpha$ anomaly near the critical point, the same as the conductivity. For a mixture of polystyrene and diethyl malonate near the liquid–liquid critical point in the absence of chemical reactions, the dielectric constant was measured [115] as a function of frequency in the range 20 KHz to 1 MHz. No anomaly was found in the dielectric relaxation time.

9.7   Conclusion

The distinctive feature of reactive systems is the existence of a homogeneous chemical mode, converting the concentration into a non-conserved parameter, which requiers a new approach in the renormalization group analysis. Milner and Martin [116] performed such analysis showing that the critical slowing down of a chemical reaction occurs for $k_H < k < k_C$, where k_C and k_H are the inverse length scales for diffusion and heat conductivity. The slowing down is governed by the strong critical index $\gamma + \alpha + \eta\nu \approx 1.37$ rather than by $\gamma \approx 1.26$ obtained in the linear theory.

Chapter 10

Phase Transitions in Moving Systems

Along the phase transitions in systems at equilibrium, there are many examples of particles undergoing phase transitions, which are carried along by the mean flow that passes through the region under study. The latter appears in different systems, such as liquids in the presence of flow [117], chemical reactions with diffusion [118], chemical waves [119], some optical systems [120] dendritic growth [121] and even population biology [122]. An additional example, which we consider in detail, is the formation of an ordered phase in the convective Ginzburg–Landau equation, which, in particular, describes the motion of vortices in superconductors [123]. This process has been observed in the superconducting films subjected to the simultaneous action of both magnetic field and bias current [124]. The vortex phase diagram in these crystals includes two distinct vortex solid phases, which are identified as ordered $(B < B_C)$ and disordered $(B > B_C)$ vortex phases, while the latter is caused by strong vortex interaction with spatial defects (pinning centers). For constant magnetic field B, the control parameter is the bias current of density J. This current will drag the vortices along the sample, helping them to destroy the disordered phase by assisting the vortices to climb the pinning barriers. The transformation of the disorder vortex phase into an ordered one in the presence of a bias current was directly observed by magneto-optical measurements of high temporal resolution, where a sharp interphase between the ordered and disordered vortex phases has been detected [124].

Vortex phase dynamics evolution in the vortex matter can be studied in the framework of the one-dimensional Ginzburg–Landau equation

$$\frac{\partial \Psi}{\partial t} + v \frac{\partial \Psi}{\partial x} = a\Psi - b\Psi^3 + \frac{\partial^2 \Psi}{\partial x^2}. \qquad (10.1)$$

The boundary conditions for Eq. (10.1) are

$$\Psi = 0 \text{ at } x = 0 \text{ and } \Psi = \text{Const as } x \to R. \qquad (10.2)$$

The new feature of our analysis is the important convection term $v\,(\partial\Psi/\partial x)$ in Eq. (10.1) which describes the motion of the vortices and contains the bias current and the threshold effect of the release of pinned vortices by the bias current.

Assume that the coefficients a and b have the simplest analytical dependence on the magnetic field H,

$$a\,(H) = \alpha\,(H_C - H)\,; \quad b\,(H) = b\,(H_C) = \beta\,. \qquad (10.3)$$

Then the free energy difference $\triangle F$ near the boundary between ordered and disordered phases, $\triangle F = a^2/b$, is equal to the magnetic energy, that is

$$\frac{H^2}{4\pi} = \frac{a^2}{b} = \frac{\alpha^2\,(H_C - H)^2}{\beta}\,. \qquad (10.4)$$

According to the Lorentz law, the velocity v is defined by the force acting on the vortex with flux quantum Φ_0,

$$v = \frac{F}{\eta} = \frac{J \times \Phi_0}{c\eta}\,, \qquad (10.5)$$

where η is the friction coefficient of the vortices [125].

Note that the experiments described in [124] were performed on $2.6 \times 0.3 \times 0.05 \ mm^3$ and $2.4 \times 0.3 \times 0.02 \ mm^3$ samples of $NbSe_2$ at temperature $5 \ K$, magnetic field $0.4 \ T$, and current density of the order of $1 \ mA/\sec$. Then, according to Eq. (10.5) the vortex velocity was on the order of $10^{-3} \ m/\sec$. The future experiments can be expanded to ac current, to systems with periodically ordered pinning centers, etc. Some of these cases are considered in what follows.

10.1 Solution of the Linearized Equation with Constant Velocity

For the stability analysis we use the linearized version of Eq. (10.1),

$$\frac{\partial \Psi}{\partial t} + v \frac{\partial \Psi}{\partial x} = a\Psi + \frac{\partial^2 \Psi}{\partial x^2}. \qquad (10.6)$$

It is convenient to introduce a new function $\Gamma(x,t)$,

$$\Psi(x,t) = \Gamma(x,t) \exp\left(\frac{vx}{2} - \frac{v^2 t}{4}\right). \qquad (10.7)$$

On substituting Eq. (10.7) into Eq. (10.6), one gets

$$\frac{\partial \Gamma}{\partial t} = \frac{\partial^2 \Gamma}{\partial x^2} + a\Gamma. \qquad (10.8)$$

The solution of Eq. (10.8) is proportional to $\exp\left[at - x^2/4t\right]$. Then, the exact solution of Eq. (10.8) will contain an exponential of the form

$$\Gamma \approx \exp\left[at - \frac{(x - vt)^2}{4t}\right]. \qquad (10.9)$$

The exact solution of the linearized equation (10.6) can be found only for a constant velocity. In contrast to immobile systems, there is more than one type of instability in moving systems. The growing mode can be shifted by flow so that locally a system remains stable, and the phase boundary is moving downstream (convective instability), while for an absolute instability the phase boundary is moving both downstream and upstream, eventually covering the entire system. The following simple arguments [126] illustrate these two possibilities.

The exact solution of the linearized equation (10.6) contains the exponential form (10.9), which describes the propagating wave packet. For each t one can find two values of x, $x = \left[(v \pm \sqrt{4a})t\right]$, which define the behavior of the two "edges" of the wave packet. For $a < 0$, there are no real x, i.e., no divergent rays, and the system is stable. For $v^2 > 4a$, both values of x have the same sign, and the solution $\Psi \neq 0$ of Eq. (10.6) is carried away with convective velocity

v (convective instability). Finally, for $v^2 < 4a$, the edges $x_{1,2}$ have different signs, i.e., the wave $\Psi \neq 0$ moves in both directions (absolute instability).

10.2 Stability Conditions for Spatially Dependent Periodic Damping

The motion of vortices is interrupted by their capture by pinning centers. One can prepare a system in such a way that pinning centers are located periodically or quasi-periodically along the system [127]. Then the convective velocity will vary periodically along the system, and the stationary version of Eq. (10.6) takes the form [128]

$$\frac{\partial^2 \Psi}{\partial x^2} - v\left[1 + b\cos\left(lx\right)\right]\frac{\partial \Psi}{\partial x} + a\Psi = 0. \qquad (10.10)$$

Equation (10.10) with periodic coefficients has a Floquet solution of the form

$$\Psi(x) = \exp(\alpha x)\psi(x) = \exp(\alpha x)\sum_{n=0}^{\infty}\left[A_n \sin\left(\frac{nlx}{2}\right) + B_n \cos\left(\frac{nlx}{2}\right)\right], \qquad (10.11)$$

where the periodic function $\psi(x)$ is expanded in the Fourier series. According to the Floquet theorem, the Floquet multiplier α must vanish at the stability boundaries. On substituting Eq. (10.11) with $\alpha = 0$ into Eq. (10.10) and comparing the harmonics in front of the sine and cosine terms, one obtains an infinite system of linear equations for A_n and B_n which have nonzero solutions if the infinite determinant $\triangle(\alpha = 0)$ vanishes, $\triangle(\alpha = 0) = 0$. One has to truncate this determinant at some n and afterwards to improve the result by taking into account the larger values of n. Leaving only terms with $n = 1$, one obtains the following equations

$$\left(a - \frac{l^2}{4} + \frac{lbv}{4}\right)A_l + \frac{lv}{2}B_l = 0,$$
$$-\frac{lv}{2}A_l + \left(a - \frac{l^2}{4} - \frac{lbv}{4}\right)B_l = 0. \qquad (10.12)$$

Equations (10.12) have nontrivial solutions if the determinant of these equations vanishes, which gives

$$b = \sqrt{4 + \frac{(4a - l^2)^2}{v^2 l^2}} \, . \tag{10.13}$$

The stability boundary (10.13) of the solution $\Psi = 0$ has a V-form in the $b - l$ plane with the stable states located inside this curve.

10.3   Stability Conditions for Space-Dependent Random Velocity

We consider now a stationary equation (10.6) with the coordinate-dependent random velocity

$$\frac{d^2 \Psi}{dx^2} - v\left[1 + \xi(x)\right] \frac{d\Psi}{dx} + a\Psi = 0 \, , \tag{10.14}$$

where the random force $\xi(x)$ is a Gaussian white noise with the correlator $<\xi(x)\xi(x_1)> = D\,\delta(x - x_1)$.
Let us rewrite Eq. (10.14)

$$L(\Psi) = v\xi \frac{d\Psi}{dx} \, , \tag{10.15}$$

where

$$L(\Psi) = \left(\frac{d^2}{dx^2} - v\frac{d}{dx} + a\right)\Psi \, . \tag{10.16}$$

In order to convert the differential equation (10.14) into an integro-differential equation we apply, following [129], the operator L^{-1} to Eq. (10.15), which gives

$$\Psi = L^{-1}\left(v\xi \frac{d\Psi}{dx}\right) . \tag{10.17}$$

Using the apparent equality $L\left[L^{-1}(\Psi)\right] = 1$, one can easily check that the integral operator L^{-1}, which is inverse to the differential operator L defined in (10.16), has the following form,

$$L^{-1}(f) = \frac{1}{a_1} \int_0^x dx_1 \exp\left[\frac{v}{2}(x - x_1)\right] \sin\left[a_1(x - x_1)\right] f(x_1)$$

$$a_1 = \sqrt{a - \frac{v^2}{4}},$$

$$(10.18)$$

i.e.,

$$\frac{d\Psi}{dx} = \frac{v}{a_1} \int_0^x dx_1 \exp\left[\frac{v}{2}(x - x_1)\right] \sin\left[a_1(x - x_1)\right] \xi(x_1) \frac{d\Psi}{dx}(x_1)$$

$$\times \left\{ a_1 \cos\left[a_1(x - x_1)\right] + \frac{v}{2} \sin\left[a_1(x - x_1)\right] \right\}. \quad (10.19)$$

On substituting Eq. (10.19) into Eq. (10.15), one obtains

$$\left(\frac{d^2}{dx^2} - v\frac{d}{dx} + a\right)\Psi(x) = \frac{v^2}{a_1} \int_0^x dx_1 \exp\left[\frac{v}{2}(x - x_1)\right] \xi(x)\xi(x_1)\frac{d\Psi}{dx}(x_1)$$

$$\times \left\{ a_1 \cos\left[a_1(x - x_1)\right] + \frac{v}{2} \sin\left[a_1(x - x_1)\right] \right\}. \quad (10.20)$$

On averaging of Eq. (10.20) for the above defined white noise $\xi(x)$, one finds

$$< \xi(x)\xi(x_1)\frac{d\Psi}{dx}(x_1) > = < \xi(x)\xi(x_1) >< \frac{d\Psi}{dx}(x_1) > = D < \frac{d\Psi}{dx}(x) >$$

$$(10.21)$$

The substitution of Eq. (10.21) into the averaged equation (10.20) shows that for white noise one gets

$$\left[\frac{d^2}{dx^2} - v(1 + vD)\frac{d}{dx} + a\right] < \Psi > = 0. \quad (10.22)$$

On comparing this equation with the stationary version of Eq. (10.20), one concludes that the existence of noise results in the renormalization of the velocity v to $v(1 + vD)$, which has to be substituted in the instability criterion $a > v^2/4$, leading to

$$a > \frac{v^2 (1 + vD)^2}{4} \tag{10.23}$$

as the condition for the appearance of an absolute instability for large values of a.

10.4 Stability Conditions for Time-Dependent Random Velocity

In the previous chapter we considered space-dependent random velocity. For the case of time-dependent random velocity, the dynamic equation has the following form

$$\frac{\partial \Psi}{\partial t} = \frac{\partial^2 \Psi}{\partial x^2} - v \left[1 + \xi(t) \right] \frac{\partial \Psi}{\partial x} + a\Psi, \tag{10.24}$$

where the random force $\xi(t)$ is a Gaussian white noise with the correlator $<\xi(t)\,\xi(t_1)> = D\,\delta(t - t_1)$.

After performing the Fourier transform, Eq. (10.24) takes the form

$$\frac{\partial \hat{\Psi}}{\partial t} = \left(a - k^2 - ikv - ikv\xi \right) \hat{\Psi}. \tag{10.25}$$

The solution of Eq. (10.25) with the initial condition $\hat{\Psi}(t = 0) = \hat{\Psi}_0$ is

$$\hat{\Psi}(k, t) = \hat{\Psi}_0 \exp[(a - k^2 - ikv - ikv)t] < \exp\left(-ikv \int_0^t \xi(\tau)\,d\tau \right) >, \tag{10.26}$$

which, after using the well-known result

$$\exp\left(-ikv \int_0^t \xi(\tau)\,d\tau \right) = \exp\left(-\frac{k^2 v^2 D_1 t}{2} \right) \tag{10.27}$$

and performing the inverse Laplace transform, gives

$$\Psi(r, t) = \exp\left[\left(a - \frac{v^2}{4(1 + v^2 D_1/2)} \right) t + \frac{2xv - x^2/t}{1 + v^2 D_1/2} \right], \tag{10.28}$$

which leads to the instability criterion

$$a > \frac{v^2}{4\left(1 + v^2 D_1/2\right)}, \tag{10.29}$$

that is the simple generalization of the condition $a > v^2/4$ in the absence of noise.

10.5 Stability Conditions for Time-Periodic Damping

Let us compare now the results obtained in the previous section with the periodically varying damping described by the equation

$$\frac{\partial \Psi}{\partial t} = \frac{\partial^2 \Psi}{\partial x^2} - v\left[1 + b\cos\left(\Omega t\right)\right]\frac{\partial \Psi}{\partial x} + a\Psi. \tag{10.30}$$

On performing calculations similar to Eqs. (10.24)–(10.28) one obtains the following solution of Eq. (10.30),

$$\Psi\left(r, t\right) = \exp\left\{\left(a - \frac{v^2}{4}\right)t + \frac{v\left[x - (bv/\Omega)\sin\left(\Omega t\right)\right]}{2}\right.$$

$$\left. - \frac{\left[x - (bv/\Omega)\sin\left(\Omega t\right)\right]^2}{4t}\right\}. \tag{10.31}$$

The instability occurs when the condition $a < v^2/4$ is satisfied, i.e., the addition of a time-periodic damping does not change the stability condition of the original equation (10.6).

10.6 Stability Conditions for a Sample of Finite Size

The finite size of a sample plays an important role in the interpretation of real experiments. As a matter of fact, for an infinite system the convective term can be simply transformed away by going to a moving frame. Returning now to the original linearized equation (10.6) one can write the solution of this equation on the finite interval $(0, L)$ with the boundary condition $\Psi = 0$ at $x = 0$ and $\Psi = \Psi_0$

at $x = L$ in the form

$$\Psi(x,t) = F_1(x) + F_2(x)\exp\left[\left(a - \frac{v^2}{4} - \frac{n^2\pi^2}{L^2}\right)t\right], \quad (10.32)$$

where $F_1(x)$ and $F_2(x)$ are some functions of x and $n = 1, 2, 3....$ Since the most rapidly growing solution corresponds to $n = 1$, a system is absolutely unstable for

$$a > \frac{v^2}{4} + \frac{\pi^2}{L^2}, \quad (10.33)$$

i.e., for large L the finite size of a sample results in a small change of the condition $a > v^2/4$.

10.7 Stability of Nonlinear Convective Ginzburg–Landau Equation

In the preceding sections, we restricted ourself to the analysis of the linearized form of the Ginzburg–Landau equation. However, nonlinearity destabilizes the linear analysis, and we now discuss the results of the analysis of the nonlinear equation (10.1). It turns out [130] that the front velocity v_f of initial droplet is

$$v_f \equiv v_1 = \frac{1}{\sqrt{3}}\left(-b + 2\sqrt{b^2 + 4a}\right) \quad \text{for } -\frac{b^2}{4} < a < \frac{3b^2}{4}, \quad (10.34)$$
$$v_f \equiv v_2 = 2\sqrt{a}... \quad \text{for } 3b^2 < 4a,$$

where v_2 is the linear marginal velocity.

For $a < -3b^2/16$, the front velocity is negative, and the droplet shrinks, implying that the original homogeneous state is stable. In contrast, for $a > -3b^2/16$, the initial droplet grows and the initial homogeneous state is nonlinearly unstable. For $a = -3b^2/16$, the homogeneous and inhomogeneous states have equal stability.

It was shown [131] that for the special case $b = 1$, in the instability region, $a > -3/16$, the instability is nonlinearly convective for $v > v_f$ since the initial droplet is eventually advected out of the system. In contrast, for $v < v_f$, the instability is absolute since the initial droplet eventually invades the system. Generalizing this argument to arbitrary values of b, one obtains from Eq. (10.34) with $v_f = v$ that the transition from convective to absolute instability occurs at

$$a = \frac{3}{16}\left(v^2 + \frac{2}{\sqrt{3}}vb - b^2\right) \quad \text{for } v < \sqrt{3}b,$$

$$a = \frac{1}{4}v^2 \quad \text{for } v > \sqrt{3}b. \tag{10.35}$$

The physical meaning of Eq. (10.35) is the following. For $v > \sqrt{3}b$, nonlinear effects are small and the absolute instability threshold remains the same as for the linear case. However, when nonlinearity effects dominate, $v < \sqrt{3}b$, the instability threshold decreases, remaining in the $a > 0$ regime.

10.8 Conclusion

Experiments dealing with the propagation of vortices in the presence of a bias current [124] open up a new area for both the experimenter and theoretician. From the theoretical point of view, the order–disorder phenomenon in vortex matter provides another example of phase transitions in moving systems. From the experimental point of view, this branch of research has opened up a chapter of studying the properties and possible new applications of superconducting films.

We considered the theoretical basis for new experiments, using ac current, films with periodically ordered pinning centers, which will introduce an additional periodic (in time and space) component to the convective velocity. In order to find the stability conditions of a disordered phase for the supercritical Ginzburg–Landau free energy, it is sufficient to perform a linear stability analysis. The stability criteria are formulated in term of the coefficients a and v in our equations, which are proportional to the magnetic field and the bias current, respectively. In the case of a constant dc current, the well-known inequalities $a < 0$, $0 < a < v^2/4$ and $a > v^2/4$ define the stable, convective unstable and absolute stable regimes, respectively. It turns out that for an additional convective velocity periodic in time these criteria remain unchanged. For an additional space-periodic component, one can point out in the amplitude-wave vector the curve (10.13), which divides the stable and unstable regions.

Noise is a factor which is inherent in all experiments. It is particularly remarkable that the component of the convective velocity random in space results in an increase of stability (see Eq. (10.23)) while the one random in time decreases the stability (see Eq. (10.29)).

Chapter 11

Random and Small World Systems

Most of the systems that we have considered so far in this book have been ordered ones, with definite well-defined spatial arrangements and interactions. In this chapter, we consider the effects of disorder or randomness. We consider first a simple example of a disordered or random system, namely that of a lattice in which there is a given probability that a bond exists between a pair of adjacent sites. Although the presence or absence of a bond is a purely geometrical property, with no thermal effects involved, this system is found to have many features similar to those of a phase transition, such as a critical fraction of bonds for percolation to occur, critical indices, and the applicability of renormalization group techniques, and so we consider it first. In fact, it is formally possible (although not very instructive) to map the bond percolation problem onto the Potts model [16] mentioned in Chapter 8. After this we consider the Ising model with random interactions or sites removed at random, and spin glasses, in which there are competing interactions. Finally, we consider small world systems, in which there are a limited number of long-range interactions between randomly chosen sites.

11.1 Percolation

The conceptually simplest example of a random system is a lattice of sites linked by bonds between adjacent sites in which a fraction x of the bonds is removed at random. In such a system, the disorder or randomness is uniquely specified by a single parameter, the fraction

x of bonds or sites that have been removed or the fraction $p = 1 - x$ of bonds that are present. For $x = 0$, which corresponds to $p = 1$, it is obvious that a continuous path of bonds exists spanning the system from one side to the other, and this will still be true if x increases and p decreases slightly. On the other hand, if $p = 0$ there is obviously no such path, and this will also be true for sufficiently small values of p. Hence it is reasonable to expect that there is a critical value p_c of p such that if $p < p_c$ then no continuous path exists that spans the system, and only isolated clusters of connected sites exist, while if $p > p_c$ an infinite cluster exists that spans the whole system. The critical value p_c is known as the bond percolation threshold, and the phenomenon is known as percolation, since if the bonds correspond to open pores in a block of material then once this fraction of pores is opened a fluid will percolate from one side of the block to the other side. It is found empirically that for a three-dimensional lattice on which the coordination number is z [132], [133],

$$p_c \approx 1.5/z\,, \qquad (11.1)$$

while for a square lattice it can be proved that $p_c = 0.5$. A related problem is that of site percolation, in which sites rather than bonds are either occupied or empty.

The percolation transition is a geometrical phase transition, in which the critical concentration p_c separates a phase of finite clusters for $p < p_c$ from a phase with $p > p_c$ in which an infinite cluster is present. This is similar to the thermal phase transitions considered in previous chapters, with the concentration of bonds p playing the role of the temperature. Just as there, the transition is characterized by the properties of clusters close to the percolation threshold. For instance, a quantity that behaves like an order parameter is the probability P_∞ that a bond belongs to an infinite cluster. By definition, $P_\infty = 0$ for $p < p_c$, while for $p > p_c$ it is found that it is related to $p - p_c$ by a power law

$$P_\infty \sim (p - p_c)^\beta\,, \qquad (11.2)$$

similar to that for the order parameter in Table 5.1. Another quantity of interest is the correlation length ξ, which is the mean distance between two sites in a finite cluster. For p close to p_c, this is given by

$$\xi \sim |p - p_c|^{-\nu}, \tag{11.3}$$

which can be compared to the behavior of the correlation length in Table 5.1. Other quantities of importance having similar types of behavior include the mean number of bonds in a finite cluster, the distribution of finite clusters of different sizes, and for $p > p_c$ the probability that a bond belongs to the infinite cluster. The problem of percolation, and especially in a continuum, is of great practical importance in many fields, such as the conductivity of a mixture of conductors and insulators and the extraction of oil from an oil-field. A comprehensive review of the percolation transition and its relationship to thermal phase transitions was presented by Essam [134].

11.2 Ising Model with Random Interactions

So far, we have considered the Ising model

$$H = -\sum_{i,j} J_{ij} S_i S_j \tag{11.4}$$

with uniquely specified values for the interactions J_{ij} between the spins on different sites. When these interactions are all ferromagnetic, $J_{ij} = J > 0$, the equilibrium state at any given temperature is uniquely defined, and if a phase transition occurs, it is between a well-ordered state and a disordered one, with a change of the symmetry of the system at the phase transition point. The Ising model with an antiferromagnetic interaction, i.e., with $J_{ij} = J < 0$, usually has similar properties. However, for certain types of lattice, there is not a unique equilibrium state at low temperatures because of frustration. The simplest example of frustration is the antiferromagnetic triangular lattice in two dimensions. Let us consider three mutually adjacent atoms, A, B, and C. If the spin at site A is up, the interaction favors the spins at sites B and C to both be down, while the interaction between the spins at B and C wants them to be of opposite sign. We now examine the Ising model for random systems, in which the values of J_{ij} are random variables with a probability distribution $P(J_{ij})$, rather than the random values of the spin considered in Section 8.3.

A simple type of random system is one with diluted ferromagnetic bonds,

$$P(J_{ij}) = x\delta(J_{ij}) + (1 - x)\delta(J_{ij} - J), \qquad (11.5)$$

for (i, j) nearest neighbors and zero otherwise, which is the usual Ising model with a fraction x of the bonds removed. A related system is the ferromagnetic system with a fraction x of the sites removed,

$$H = -J \sum_{i, j} c_i c_j S_i S_j, \qquad (11.6)$$

with $P(c_i) = x\delta(c) + (1 - x)\delta(c - 1)$. In these cases, for sufficiently low values of the dilution parameter x the system exhibits a phase transition from a paramagnetic state to a ferromagnetic one as the temperature is reduced. However, for sufficiently large values of x, there is no ferromagnetic phase but only a transition from the paramagnetic phase to the Griffiths phase [135], in which there are clusters of ordered spins but not all the clusters are ordered in the same direction. The above results are connected with percolation theory. For sufficiently large values of x, which correspond to sufficiently small values of p in the percolation problem, there is no connected cluster of spins spanning the system but only isolated clusters of connected spins, and at sufficiently low temperatures the spins in each cluster will be ordered, but those for different clusters will be ordered in arbitrary directions.

Another type of system is one in which the interactions are not only of random strength but also of random sign, for instance with $P(J_{ij})$ having a Gaussian distribution,

$$P(J_{ij}) = (2\pi\Delta)^{-\frac{1}{2}} \exp[-J_{ij}^2/(2\Delta)] \qquad (11.7)$$

or a bimodal distribution

$$P(J_{ij}) = x\delta(J_{ij} + A) + (1 - x)\delta(J_{ij} - A), \qquad (11.8)$$

frequently studied for the symmetric case $x = \frac{1}{2}$. In systems with such combinations of ferromagnetic and anti-ferromagnetic interactions, frustration of the spins occurs, similar to that mentioned above for the antiferromagnetic Ising model on a triangular lattice. As a result, these systems, which are known as spin glasses, have no unique ground state.

11.3 Spin Glasses

There are a variety of magnetic systems in nature that behave as spin glasses. One type is that of magnetic ions such as iron or cobalt inserted as substitutional impurities at random sites in a lattice of non-magnetic metal ions such as copper. These systems have frozen disorder, similar to that of a glass at low temperatures, and the interactions between the spins on the magnetic ions are indirect, i.e., they are mediated by the spins of the electrons attached to the host atoms or by free electrons. In the latter case, there is an effective magnetic interaction of the Ruderman–Kittel form [136]

$$\phi(r) = \phi_0 \sin(k_F r)/(k_F r)^3 \,, \tag{11.9}$$

where k_F is the wave number of electrons on the Fermi surface, so that the sign of this interaction varies with the distance r between the magnetic ions.

A major difference exists between ferromagnetic system and spin glasses with respect to their symmetry properties. In ferromagnetic materials, when the temperature is raised through T_c the equilibrium state of the system changes to the paramagnetic one, with a sudden change in the magnetic moment M and symmetry breaking. Moreover, in ferromagnetic systems such as the original Ising model, at temperatures $T < T_c$ the system is ferromagnetic, with all the spins pointing in the same direction, and there is degeneracy between the states with all spins up and all spins down. In the thermodynamic limit, these two states are separated by an infinite potential barrier, so that no transition between them can occur. As a result, in the ferromagnetic state not all states in phase space are allowed. Since a system is ergodic only when it is free to sample all points in phase space, this means that the ergodicity is broken at T_c. Incidentally, the degeneracy of the ferromagnetic state can be removed (in order to produce a unique ground state) by applying a small magnetic field or by introducing suitable boundary conditions. Thus, for ferromagnetic systems, both symmetry and ergodicity are broken at T_c. In spin glasses, by contrast, while there is a critical temperature T_c, the mean magnetic moment $< M >$ is zero both below and above it, and the transition is characterized by a cusp-like singularity of the susceptibility at T_c [137]. Below the critical temperature there are many competing ground states, separated by infinite barriers. This means that for spin glasses the ergodicity is broken at T_c (as a result of frustration), but this broken ergodicity is not accompanied by broken symmetry

In all random systems, the macroscopic properties are those of a configurational average of the system, which we denote by $< \cdots >$, i.e., the property has to be calculated separately for each possible accessible configuration of the system and an average taken (with the appropriate weights of the probability of that configuration being found) over all the possible configurations. In general, there are two types of random system, annealed or quenched, depending on the rate of change of the disorder, for instance as a result of the mobility of magnetic ions in spin glasses. In annealed systems this rate is fast, so that one requires the average of the positions of the impurities which enter the partition function, and one has to average the partition function $< Z >$. On the other hand, in quenched systems this rate is slow and the disorder is essentially fixed, so that one has to minimize the free energy $F = -kT \ln Z$ individually for each possible configuration, and so the important quantity is $< \ln Z >$. In spin glasses the disorder is quenched, so that for the macroscopic system one has to average over the free energy for different configurations rather than over the partition function. This is the main problem in the theoretical analysis of spin-glasses.

There are two well-known models for spin glasses, the "short-range" Edwards–Anderson and the "infinite-range" Sherrington–Kirkpatrick models, which overcome the difficulty stemming from the nonlinearity of the logarithm function appearing in the expression for the quenched free energy by the so-called replica method [138] based on the self-evident identity,

$$\ln Z = \lim_{n \to 0} \frac{Z^n - 1}{n}. \tag{11.10}$$

Equation (11.10) means that one can calculate the quenched average free energy by creating multiple copies of the system, averaging them, and then letting the number of copies tend to zero. The calculation of Z^n is similar to that performed in Chapter 8 for the spherical model, but with many delicate points, such as the meaning of the limit $n \to 0$, its interchange with the thermodynamic limit, etc.

There are a number of problems from different fields that can be related to the spin glass problem, and we will mention two of them. The theory of spin glasses can be used to derive computer algorithms for the solution of optimization problems, such as that of the travelling salesman [139] (who has a large number of possible routes, and if he spends all his time trying to work out the best one he will never sell anything). Another problem related to spin glasses is that of neurons in the brain, which can be in one of two states ("firing" or "non-firing"), just like the spins in the Ising model. An appropriate model to describe the possible patterns of neural connections was introduced by Hopfield [140], and has led to an enormous amount of work on neural networks, with applications to understanding such functions of the brain as learning, and to attempts to mimic them so as to obtain self-correcting systems.

11.4 Small World Systems

We are living in a large world, and already in 1999 the population of the Earth exceeded six billion people. On the other hand, if we suddenly meet our neighbor somewhere in the Amazon jungle, we exclaim: "This is a small world!". The simple "large–small" games can be illustrated by the Bacon and Erdos numbers. The famous actor Kevin Bacon acted in many movies with different actors, who can thereby be characterized by their different Bacon numbers, where the Bacon number of an actor is one if he appeared in the same film as Kevin Bacon (1,472 people), two if he appeared in the same film as an actor with Bacon number one (110,315 people), etc. The number of people with Bacon number three is 260,123, and so on. In the world of science, a similar process is connected with the name of a legendary mathematician Paul Erdos, who published more than 1,500 scientific papers with 458 collaborators, each of whom has Erdos number one. Each of the additional 4,500 scientists who collaborated with the latter have the Erdos number two, and so on. It turns out that majority of scientists publishing articles in the last sixty years have quite small Erdos numbers. The author of this book has Erdos number three, having worked with G. Weiss who worked with J. Gillis who worked with P. Erdos.

However the first real, though quite bizarre, experiment was carried out by the psychologist Milgram [141], who tried to find the "distance" between two randomly chosen American citizens, which is defined as the number of acquaintances needed to connect these two individuals. The results that one expects from such an experiment are not at all clear. On the one hand, each person knows (calling them by the first name) about one hundred individuals, each one of those knows another hundred, etc., so 280 million Americans will be covered already in five steps $(280,000,000 \ll (100)^5)$. This simple argument is obviously wrong, since generally friends of my friends are likely also to be my friends. The opposite point of view is that an infinite number of steps is needed to find a connection between, say, a single homeless person and the president of the USA. The surprising result of Milgram's experiment was that everyone in the world is connected to everyone else through a chain of at most six mutual acquaintances. The legendary phrase "six degrees of separation" became an integral part of our "small world".

However, this subject was given a firm scientific basis, and the concept of "small world" became a normal scientific term, only thirty years later in the outstanding article of Watts and Strogatz [142]. They considered the transition from a regular to a random graph. In general, a graph is defined as a set of sites or vortices with connections between them which are called bonds or links. A regular graph is one

Regular   **Small-world**   **Random**

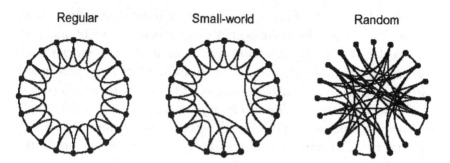

Fig. 11.1   Three types of graphs: regular, random and regular with a small number of "shortcuts" (small world). Reprinted by permission from Nature [142] copyright 2003, Macmillan Publishers Ltd.

in which each vortex is linked to its k closest neighbors (which can include nearest neighbors, nearest and next nearest neighbors, etc.), while in a random graph each vortex is linked to k arbitrary vortices. There are two important parameters which characterize the long-range and short-range properties of a graph containing $N = L^d$ sites and $\frac{1}{2}Nk$ links, where L is the linear size of the graph and d is the spatial dimension. The long-range parameter is the average minimal number of links $h(l)$ required to go from one site to another one at a distance l from it, and this is called the shortest path distance between the sites. The l-dependence of this shortest path distance is quite different for regular and random graphs. While for a regular graph, it is obvious that $h(l) \sim l$, for a random graph $h(l) \sim \ln(l)$. For a simple argument to support this result, we note that each site has on the average k sites linked to it, and each of these is also linked to k sites, so that a given site has k^2 sites with $h = 2$, etc. Thus after h steps one can reach a site located a distance l away, where $k^h = l$ or $h = \ln(l)/\ln(k)$. One can also define a clustering coefficient c which is the average fraction of neighboring sites linked to a given site. This parameter defines a short-range order, and is obviously close to unity for a regular graph and very small for random graphs. Hence, for a regular graph h and c are both large, while for a random graph they are both much smaller.

A new type of graph, intermediate between regular and random, as shown in Fig. 11.1, was introduced by Watts and Strogatz in the following way [142]. They started from a regular one-dimensional graph, where each site is connected to its k nearest neighbors. They then replaced some of these connections by links between randomly chosen pairs of sites ("shortcuts"), with probability p per link, which we will call the randomizing number. This leads to an average of $pkN/2$ such shortcuts. The two limit cases are $p = 0$ (regular graph) and $p = 1$ (random graph). For all intermediate values, $0 < p < 1$, the graphs are partially random (disordered). Strogatz and Watts founded that already for a small value of p, when the graph is still locally ordered having only a small number of "shortcuts", the graph exhibits a different behavior, namely $h(l) \sim \ln(l)$, as in random graphs, but c remains close to unity, similar to regular graphs. Such a behavior is known as "small world" [142]. The drop in h from $h \sim l$ to $h \sim \ln(l)$ is connected with the small number of shortcuts which bring together previously distant sites.

Now that we have three types of graph, regular ("large world"), disordered ("small world") and random, we have to decide which type is most common in real life. To do this, it is convenient to change our terminology, and assume that the sites in the graphs are occupied by some objects, and that links correspond to interactions between these objects. These new systems will be called networks. The concept of networks is very general, and incorporates such systems as interacting atoms and galaxies, different groups of people, sites on the Internet, etc. In their original article Watts and Strogatz considered three different networks (a sociological system of film actors, a biological system of the worm's neural network, and the man-made power grid in the western US), and found that all of them exhibit the small world phenomena. A more recent development is that many other networks belong to the small world category (see Watts' book [143] and some recent reviews [144], [145]). In addition to the topological characteristics h and c of the network, some new parameters (efficiency and cost, geographic distances [146], etc.) have been introduced for the description of real networks.

11.5 Evolving Graphs

So far we have examined graphs with a fixed number of sites and links. However, most real systems are "living", i.e., their size L changes with time. Accordingly, let us consider graphs which are characterized by three parameters, the number k of closest neighbors, the randomizing number p, and the total number of sites L^d. In the previous section we considered the p-dependence of two characteristic parameters, the path length $h(l)$ and the cluster coefficient c, provided that k and L remained constant. As we saw, the shortest path distance $h(l)$ between two sites a distance l apart is quite different for regular and random graphs, being proportional to l and $\ln(l)$, respectively. The question arises as to the conditions for a transition from one regime to another to take place, or in other words how large the disorder has to be for a transition to take place from a regular lattice to a random one. Roughly speaking, a graph is regular when the fraction of random links N_{rl} is small, $N_{rl} \ll 1$, and it is random when this fraction is large, $1 - N_{rl} \ll 1$. The total number of links is $\frac{1}{2}kL^d$, where d is the spatial dimension, and the probability of a random link is p, so that the average fraction of random links N_{rl} is given by $N_{rl} = \frac{1}{2}kL^d p$. Since N_{rl} is small for regular graphs and close to unity for random ones, it seems natural to assume that the transition occurs at some "crossover" size L^* such that

$$N_{rl} = \frac{1}{2}kL^{*d}p = \frac{1}{2}, \qquad (11.11)$$

i.e., the graph with given k and p is regular when $L < L^*$ and random when $L > L^*$. The heuristic arguments which lead to Eq. (11.11) are supported by the renormalization group analysis [147]. For a one-dimensional system, it follows that

$$L^* \sim \frac{1}{p}. \qquad (11.12)$$

We note that the transition from a regular to a random graph occurs at some value L^* of L no matter how small the randomizing number p is. Moreover, an analogy exists between this "geometric" phase transition and the "physical" phase transitions considered in previous chapters, with L and p playing the role of the correlation length and the dimensionless temperature, respectively. Later on we will compare the characteristic length L^* with the correlation length ξ.

11.6 Phase Transitions in Small World Systems

In the previous section we considered the "geometric" phase transition controlling the transition from regular to random graphs. There are, however, also "physical" phase transitions, if the small world defines the underlying topology for physical objects located on sites of these graphs, with the links defining the interactions between these objects. In such a way we obtain physical objects with random interactions. A simple example is provided by the Ising model applied to linear polymers, where the short-range interactions between neighboring monomers in the chain are supplemented by the random interactions between monomers that are close in space (and not along the chain). This example demands a slight change of the original model [142] in which the shortcuts do not replace the short-range interactions but rather are added to them, i.e., no links are removed from the regular lattice but extra links are introduced.

Let us recall the main results of the regular one-dimensional Ising model considered in Chapter 3. Such a system with N sites has no phase transition at finite temperatures in the thermodynamic limit $N \to \infty$. However, a mean field type phase transition is found in the Kac model [52] where, in addition to the short-range interactions, one adds a large (on the order of N^2) number of weak interactions between all pairs of spins. The ferromagnetic one-dimensional Ising model has been analyzed analytically and numerically in small world systems. It is found from analytical calculations [148], [149], that a mean field transition occurs at a finite temperature $T_c(p)$ for any $p > 0$ provided that the number of sites is large enough, and numerical simulations show that the critical indices have their mean field values [150]. Close to $p = 0$ the transition temperature T_c goes to zero as $|\ln(p)|^{-1}$. Hence, in order to obtain the mean field phase transition in a one-dimensional Ising system there is no need for long-range interactions between all the spins, and it is enough to have a comparatively small number of randomly chosen long-range interactions. We note that a mean field type behavior has also been found by Monte Carlo simulations for the $x-y$ model based on two-dimensional small world graphs [151].

The problem becomes more interesting in two and three dimensions where phase transitions already occur in the ordered $(p = 0)$ systems, and one expects that the appearance of a small amount of disorder $(p \ll 1)$ will result in a change of the universality class of this transition from that of the Ising model for $p = 0$ to that of the mean field for $p > 0$. If this change occurs at finite p, the question arises as to the dependence of the critical indices on p in the transition regime. It turns out, however, that the transition to the mean-field type behavior probably occurs for all $p > 0$, provided that the number of sites is large enough, although numerical simulations [151], [152] have been restricted to $p \geq 10^{-3}$. If this last statement is correct, the question arises as to the dependence on p of the transition temperature T_c for small p. An elegant answer to this question has been found by Herrero [152]. Let us compare the two characteristic lengths appearing in our problem, the characteristic length L^* which defines the transition from regular (large world) to random (small world) behavior, and the correlation length ξ of the "physical" phase transition. When ξ is smaller than L^*, the system behaves as a regular lattice. However, as the temperature decreases ξ becomes larger than L^*, and so the random interactions become important and give rise to the mean field behavior. Thus the transition between the regular lattice (Ising type behavior) and the small world (mean field behavior) occurs for $\xi \approx L^*$. For the one-dimensional Ising model the correlation length at low temperature is given by $\xi \sim \exp[2J/(kT)]$ [67], and for $L^* \sim p^{-1}$, the condition $\xi \sim L^*$ results in $T_c \sim |\ln(p)|^{-1}$ in agreement with [149]. For higher dimensions, $\xi \sim |T - T_{c,p=0}|^{-\nu}$ and $L^* \sim p^{-1/d}$. Thus, the change in the temperature of the phase transition for a non-zero randomizing parameter p is given, for small p, as

$$T_c - T_{c,p=0} \sim p^{\frac{1}{\nu d}}. \tag{11.13}$$

11.7 Conclusion

Although systems with random interactions have been studied for many years, especially in the framework of spin glass theory, a fresh point of view was introduced a few years ago with the discovery of small world phenomena. It turns out that most real "many-body" natural and human-made systems exhibit small world properties, having a short-range structure like that of ordered systems and a long-range structure similar to that of random systems. These unusual properties lead to many important applications in science and sociology [144], [145].

Chapter 12

Self-Organized Criticality

Many-particle systems can be divided into two classes. On the one hand, one can start with a superficially simple system such as an atom, and then consider more and more complex systems as one probes further and further into its structure, electrons and a nucleus, the nucleus containing protons and neutrons, these being composed of quarks, and so on. On the other hand, one can consider large systems that are inherently complex because they contain many objects, with very many degrees of freedom and a large number of connections between them. Just as in the previous chapter, these systems can be described in terms of networks of sites and bonds. Such a network can refer not only to the atoms or molecules in a solid or a fluid but also to groups of people and their interactions, competing companies on the stock market, the sites and links of the Internet, and numerous other systems. In our analysis we consider those properties which are independent of the type of objects located at the sites and connected by the bonds, but rather depend on the topological properties of networks. Such properties are very appropriate to phase transitions, where as we saw in Chapter 8 the universality classes do not depend on the detailed properties of the objects being associated with the infinite correlation length.

In the previous chapter we classified networks by the parameters h and c. Another way of doing so is to introduce the probability $P(l) \sim l^{-\delta}$ that two sites at a distance l are connected. If $\delta = 0$, $P(l)$ is constant, so that links of all sizes are equally probable, as in the random graphs discussed in the previous chapter. On the other hand, if δ is very large then long-range links are restricted, and one effectively obtains a regular graph with only short-range links. For intermediate values of δ, one gets the original Watts–Strogarz model of a small world [142]. It turns out [153] that for $\delta < \delta_1$ the system behaves as a random network, for $\delta_1 < \delta < \delta_2$ the behavior is of the small world type, and for $\delta > \delta_2$ it behaves as a regular network, while for a one-dimensional system $\delta_1 = 1$ and $\delta_2 = 2$.

For the probability of a connection between two sites, we introduced the power law distribution $P(l) \sim l^{-\delta}$. In general, for independent random events with a finite mean a and variance σ, the central limit theorem states that a property x will have the Gaussian distribution,

$$P(x) = (2\pi\sigma)^{-\frac{1}{2}} \exp[-(x-a)^2/2\sigma]. \qquad (12.1)$$

This bell-shape curve distribution is probably the most common type of figure in the scientific literature. As we have seen in the previous chapters, in some special cases the Gaussian distribution is replaced by a power-law one. The two examples that we examined previously are the spatial dependence of the correlation function at critical points and of the correlation length in the low temperature region of the x–y model. In contrast to these special cases of systems in equilibrium, which only apply to some special temperature region, there is a wide class of non-equilibrium phenomena which are of the power law type. Since these systems evolve spontaneously into critical states because of their internal dynamics, with no dependence on the value of an external parameter, their behavior is called self-organized criticality (SOC). An alternative name for such systems is scale-free systems, since the power law distribution is scale-invariant and does not involve a characteristic length scale. This situation is similar to that which occurs close to the critical points in equilibrium systems, where the short-range lengths are of no importance and the only characteristic length is the correlation length. This diverges at the critical point, so that there is no characteristic length at this point.

12.1 Power-Law Distributions

In considering power-law distributions, we start with some old empirical laws. One of the oldest examples (1897) is the Pareto distribution [154] describing the distribution of large wealth w in different societies. It turns out that the number of people $N(w)$ having wealth w has a form $N(w) \sim w^{-\alpha}$ with α between 2 and 3. Very recently [155], the effects of using fiscal policy to change the value of α have been considered, and some surprising results have been found with regard to the probability of an individual becoming very wealthy (wealth condensation). A similar power-law distribution was found by Guttenberg and Richter [156] in 1954 for the number $N(E)$ of earthquakes which release a certain amount of energy E. This dependence has the form $N(E) \sim E^{-\alpha}$, with $\alpha = 1.5$ independent of the geographic region examined. Finally, in the 1930's, Zipf found [157] the power-law distributions of towns according to their populations and of the frequency of appearance of different words in the English language. There are many more systems with power-law distributions, such as climate fluctuations, cloud formation, rainfall, volcanic eruptions, forest fires, commodity prices, scientific collaborations, citations, the structure of the Internet web and traffic jams. One of the most recently found and exotic examples is the power-law distribution of the wealth of the inhabitants of the ancient Egyptian city Akhetaten, based on the excavations and on the assumption that wealth was proportional to the house area [158].

The power-law distribution, unlike the Gaussian distribution, has the form of a straight line in a logarithmic plot and has no distinctive value. This feature has some interesting consequences, of which two are of special importance. First of all, it is meaningless to define a typical income or the average strength of an earthquake, which would correspond to the peak of a Gaussian distribution. Secondly, and of greater importance, the straight line logarithmic distribution suggests that the same mechanism is responsible for earthquakes of all sizes, including the largest ones. Such large values are often regarded as exceptions, or catastrophic, but the power-law distribution shows that although they are rare in occurrence, they are expected to occur occasionally, and do not require any special explanation. This may be counter to our physical intuition, in which we are used to systems in thermal equilibrium within deep potential minima, where the response is proportional to the disturbance. In that case, small disturbances lead to small consequences and some special mechanism is required to produce large effects. However, there are occasional exceptions even for systems in equilibrium. One example is the divergent susceptibility at the critical point, which leads to a small disturbance producing a large response near this point. In this manner, the gravitational field, which in general is negligible for all laboratory experiments, becomes important near the liquid–gas critical point. Because of the infinite compressibility $\partial\rho/\partial p$ at the critical point, even the small change in pressure δp along a vertical tube 10 cm long leads to a large difference $\delta\rho \sim (\partial\rho/\partial p)\delta p$ between the fluid's density at the top and bottom of the tube ("gravity effect" [159]), which can be as large as 10%. Similarly, in systems driven at a frequency close to the system's resonance frequency a small force can produce a catastrophic response, as happened in the collapse of the Takoma Bridge on 7 November 1940.

A basic difference between the above systems and those that we are considering now is that the latter are not in thermodynamic equilibrium. One can imagine, for example, a shallow valley in the potential landscape, in which case only a small stimulus is required to lift the system out of this valley, and it can then fall into another potential valley of arbitrary depth. A simple example of this is provided by a man standing on a mountain. If he is at the bottom of the mountain, a small push will have very little effect. If he is standing on a ridge which has a wider ridge slightly below it, then a small push will at most cause him to fall a short distance. However, if he is standing at the top of the mountain with a sheer drop to down below, then a small push away from the edge will have little effect, but a small push over the edge will have catastrophic results. Let us give two illustrative examples of such catastrophic effects. While one can easily find the obvious reason for the crash of New York stock exchange after 11th September 2001, it is very hard to find a special reason for the "black Monday" of 19th October 1987, the greatest one-day crash in history, when that market lost 20% of its value in one day. The second example concerns the disappearance of dinosaurs. It is possible that this was due to the collision of Earth with an asteroid, or due to a drastic change of the climate, or due to a huge volcanic eruption. However, there may be no need for a special explanation, as this phenomenon — the disappearance of dinosaurs — could have happened due to the internal properties of the ecosystem.

12.2 Sand Piles

The paradigm for power-law distributions, and the one which was the subject of the first serious theoretical studies, is provided by sand piles. As grains of sand are dropped one by one, slowly and uniformly, onto a table, at first they stay where they land, forming an initial thin layer, and then grains will accumulate on top of one another creating a pile. However, this pile will not grow taller forever. At some stage, the sand pile will enter a state of self-organized criticality (SOC), with a slope of 34°–37°, and then a qualitative change will occur. A new falling grain has the following three possibilities:

a) It can fall to the bottom of the pile, and so enlarge the pile's base.

b) It can stick to the side of the pile at some point, thereby increasing slightly the angle of slope of the pile.

c) It may increase the angle of slope above a critical value and so trigger an avalanche in which S grains fall to the bottom of the pile, some of which may have been located nowhere near the place where the grain which started the avalanche fell. This happens because in the SOC state the whole pile has become connected, so that even a single grain (small disturbance) in one part of the pile may trigger a large avalanche (catastrophic event) anywhere else. In the SOC state the global dynamics controls the behavior of the sand pile as a whole rather than that of the individual grains. In this SOC state the number $N(S)$ of avalanches involving S grains is found to have the power law distribution $N(S) \sim S^{-\alpha}$ with $\alpha \approx 1$. It is intuitively clear that there is no typical size of an avalanche, and there are no obvious precursors of a huge, catastrophic avalanche. Since each grain "knows" only its short-range environment, it does not expect to be involved in an avalanche, and no predictions of the catastrophic events can be made.

For a mathematical model of such a process, one describes the system as a box which is divided into cells, and assumes that there is an upper limit to the number of particles that can accumulate in a single cell. When a cell contains this maximum permissible number of particles, the arrival of an additional particle will cause some particles to leave it for other cells. If one or more of these particles move to cells which are located near the boundaries of the box and already contain the maximum number of particles, then more and more particles will leave the system creating thereby an avalanche.

A similar model can be used for study of forest fires. For the model of a forest, the boxes consist of cells with trees inside them. Then, one of the trees is set on fire, and one studies how many trees will be consumed before the fire extinguishes itself. A study of this system permits the planning of how to plant forests in such a way as to minimize the risks of large fires. It may be useful to initiate occasionally small fires so as to reduce the connections between different groups of trees (cells), and keep the whole system below the critical SOC level when fires of arbitrary size (including huge avalanches) become possible. It is possible that the wrong policy of keeping the ecosystem in its virgin form, and so suppressing the small fires, was responsible for not preventing the huge 1988 fire in Yellowstone National Park in USA. The forest fire professionals now recognize that the best way to prevent the largest fires is to allow small and medium size fires to burn.

12.3 Distribution of Links in Networks

As in the previous chapter, let us consider a network consisting of a large number N of sites each having a number k links to other sites in the network. Different sites may have different number of links, and so we introduce the probability distribution of the number k of links, $P(k)$, where k_i is the number of links (also called connectivities) of site i. For scale free systems $P(k)$ is the power law function, $P(k) \sim k^{-\gamma}$.

In order to show how a power-law distribution can appear, we associate with each site i a random number $x_i \in (0, \infty)$, which we call the rank of site i, and use it to measure the importance of this site [160]. Let $f(x)$ be the probability distribution of the ranks, which for scale-free systems we take to be a power-law distribution similar to those considered previously in this chapter. In addition, we assume that the probability that a link exists between a pair of sites i and j is $\psi(x_i, x_j)$. Incidentally, for random graphs, this probability is the same for all sites coupled, and $\psi(x_i, x_j)$ is a constant, $\psi(x_i, x_j) = b$.

The mean connectivity of a site of rank x is

$$< k(x) > = \int_0^\infty \psi(x, y) f(y) dy \equiv NF(x), \qquad (12.2)$$

so that $x = F^{-1} \left(\frac{<k>}{N} \right)$. On the other hand, by definition

$$< k > = \int_0^\infty kP(k) dk. \qquad (12.3)$$

On combining Eqs. (12.2) and (12.3), one obtains, using $P(k) = f(x) dx/dk$,

$$P(k) = f \left[F^{-1} \left(\frac{<k>}{N} \right) \right] \frac{d}{dk} \left[F^{-1} \left(\frac{<k>}{N} \right) \right]. \qquad (12.4)$$

As a special example, we consider $\psi(x_i, x_j) = x_i x_j / x_{max}^2$, where x_{max} is the largest rank in the network. Then,

$$< k(x) > = \frac{Nx}{x_{max}^2} \int_0^\infty y f(y) dy = N \frac{x < x >}{x_{max}^2}, \qquad (12.5)$$

and

$$P(k) = \frac{x_{max}^2}{N < x >} f \left(\frac{x_{max}^2}{N < x >} k \right). \qquad (12.6)$$

One sees from Eq. (12.6), that the distribution of connectivities $P(k)$ is completely determined by the distribution of ranks $f(x)$, so that if $f(x)$ is a power-law distribution, then so is the distribution $P(k)$ of connectivities.

12.4   Dynamics of Networks

So far we have considered only static networks. However, most networks are not static, as the number of sites can change with time, new sites and links can appear, and some of the old ones may disappear. For instance, in the world wide web of the Internet, new sites are continually being created, as are new links between sites, while some sites may also disappear. Similarly, in the financial world, new businesses are continually being created and others liquidated, while assets are also transferred from one firm to another.

The main assumption which leads to a power-law distribution of links is that of preferential attachment [161], which means that new links prefer to be connected to those old sites that already have many connections (which corresponds to "rich becomes richer"). Let us consider a model system that initially contains m_0 bonds and n_0 sites. At each of the time steps $t_1, t_2, \ldots, t_N$, where we choose $t_r = r$ ($r = 1, 2, 3, \ldots$), one site with m bonds is added to the system so that after t time steps the network has $N = n_0 + t$ sites and $m_0 + mt$ bonds. According to the assumption of preferential attachment, the increase of the connectivity k_i on site i is chosen to be proportional to k_i itself,

$$\frac{\partial k_i}{\partial t} = A \frac{k_i}{\sum\limits_j k_j}, \qquad (12.7)$$

where the normalization parameter $\sum\limits_j k_j$ determines the total number of bonds, and at long times $\sum\limits_j k_j \simeq 2mt$. The factor two in this formula appears because each bond links two sites. On summing Eq. (12.7) over all i, we find that the coefficient A determines the change of the total number of connectivities per unit time, i.e., $A = m$. The substitution of this result in Eq. (12.7) and solution of the resultant equation leads to

$$k_i = m \left(\frac{t}{t_i} \right)^{\frac{1}{2}} \quad \text{or} \quad \frac{dt_i}{dk_i} = -\frac{2m^2 t}{k_i^3}. \qquad (12.8)$$

The distribution of links $P(k)$ is related to the (homogeneous) distribution of the times t_i when the sites were created, $P(t_i) = 1/t_i$, by

$$P(k)dk = -P(t_i)dt_i \,. \tag{12.9}$$

Here, the minus sign is connected with the fact that the older sites (with small t_i) have larger connectivity ("rich become richer"). On combining Eqs. (12.8) with (12.9), we finally obtain

$$P(k) = \frac{2m}{k^3} \,. \tag{12.10}$$

According to Eq. (12.7), the increase of the connectivity of a given site is fully determined by the present value of its connectivity, and this gives an obvious advantage to the "old" sites as compared to the "new" ones which only recently jointed the network. However, in addition to its "age", the "quality" of a site should influence its future development. To extend the model of equation (12.7) in this respect, we now introduce a fitness parameter γ_i for site i, with $0 \le \gamma_i \le 1$, which describes how capable it is of receiving links. The increase of the connectivity depends now not only on the present value of the connectivity k_i of the site i, as in Eq. (12.7), but also on its fitness. Accordingly we assume that

$$\frac{dk_i}{dt} = m \frac{\gamma_i k_i}{\sum\limits_j (\gamma_j k_j)} \,. \tag{12.11}$$

An interesting analogy exists [162] between networks with the dynamics described by (12.11) and the equilibrium Bose gas considered in Chapter 7. In order to establish this, we associate with the fitness γ_i an energy ε_i through the relation

$$\varepsilon_i = -T \ln(\gamma_i) \,, \tag{12.12}$$

so that $0 < \gamma_i < 1$ corresponds to $\infty > \varepsilon_i > 0$, where in this section we have set the Boltzmann constant equal to unity. Let a link between the two sites i and j (which have fitnesses γ_i and γ_j or energies ε_i and ε_j) be replaced by two non-interacting particles. Then the addition of a new site l with m links to the network means the addition of $2m$ particles, half of which have the energies ε_l, and the other half of which are distributed between other energy levels corresponding to the endpoints of the new links. Instead of the connectivity of a site depending only on time, as in Eq. (12.7), the connectivity k_i on site i in the generalized theory will also be a function of the number of links (particles) which the site has at time t, $k_i = k_i(\varepsilon_i, t, t_i)$. Equation (12.11) will take the following form

$$\frac{\partial k_i(\varepsilon_i, t, t_i)}{\partial t} = m \frac{\exp(-\varepsilon_i/T) k_i(\varepsilon_i, t, t_i)}{\sum_j \exp(-\varepsilon_j/T) k_j(\varepsilon_j, t, t_j)} \, . \tag{12.13}$$

By analogy with Eq. (12.8), we look for a solution of Eq. (12.13) of the form

$$k_i = m \left(\frac{t}{t_i} \right)^{f(\epsilon_i)} . \tag{12.14}$$

One can rewrite the "partition function" $Z(t)$ in the denominator of Eq. (12.13) as

$$Z(t) = \int_0^\infty d\varepsilon g(\varepsilon) \int_1^t dt' \exp\left(-\frac{\varepsilon}{T}\right) k(\varepsilon, t, t') , \tag{12.15}$$

where $g(\varepsilon)$ is the distribution of energies obtained by Eq. (12.12) from the distribution of fitnesses $\phi(\gamma_i)$. On substituting Eq. (12.14) into Eq. (12.15) one can perform the integration over t_i, to obtain

$$Z(t) = mt \int_0^\infty \frac{d\varepsilon \, g(\varepsilon) \exp(-\varepsilon/T)}{1 - f(\varepsilon)} \, . \tag{12.16}$$

We now introduce the "chemical potential" μ as

$$\exp\left(-\frac{\mu}{T}\right) \equiv \lim_{t \to \infty} \frac{Z(t)}{mt} \, . \tag{12.17}$$

From Eqs. (12.17), (12.13) and (12.14), one then finds that

$$f(\varepsilon) = \exp\left(-\frac{\varepsilon - \mu}{T}\right) , \tag{12.18}$$

and obtains the following equation for μ,

$$\int_0^\infty \frac{d\varepsilon\, g(\varepsilon)}{\exp[(\varepsilon - \mu)/T] - 1} = 1 \,. \qquad (12.19)$$

Equation (12.19) is nothing other than Eq. (7.22) for the chemical potential of a Bose gas. The mapping of our problem to that of a Bose gas means that the typical property of a Bose gas, Bose–Einstein condensation, must also occur in an evolving network. Therefore, one predicts, in addition to the usual self-organized criticality with preferential attachment ("rich become richer"), the appearance of a new phase, analogous to the Bose condensate, where all the new sites will correspond to $\varepsilon = 0$, which according to Eq. (12.12) corresponds to the site with the highest fitness $\gamma = 1$. It is just this site that will have links to all new sites ("winner takes all").

In addition to fitness, there are some other generalizations of the main dynamic equation (12.7).

a) If one assumes in the phenomenological expression (12.7), $\partial k_i/\partial t \sim k^\beta$ with $\beta \neq 1$, one obtains [163] an exponential distribution for $P(k)$ if $\beta < 1$, the limit situation of "winner takes all" if $\beta > 1$, and only for $\beta = 1$ does $P(k)$ exhibit power law behavior.

b) The power-law distribution function was obtained from the assumption of preferential attachment (12.7). On the other hand, in random graphs and small world systems, a new site randomly creates new links with each one of the existing sites. In contrast to the preferential attachment, this assumption leads to an exponential form of the distribution function. It can be assumed [164] that a realistic network grows in time according to an attachment rule that is neither completely preferential nor completely random, i.e.,

$$\frac{\partial k_i}{\partial t} = \frac{(1 - p)k_i + p}{\sum_j [(1 - p)k_j + p]} \,, \qquad (12.20)$$

where $0 \leq p \leq 1$ is a parameter characterizing the relative weights of the deterministic and random attachments. An analysis similar to

Eqs. (12.7)–(12.10) leads to the following result

$$P(k) \sim \left(\frac{k/m + b}{1 + b} \right)^{-\gamma}, \qquad \gamma = 3 + b, \qquad b = \frac{p}{m(1 - p)}, \qquad (12.21)$$

where, as previously, m is the number of new links added at each time step. The power-law behavior for scale free networks is recovered from Eq. (12.21) if $p \to 0$, while the exponential distribution dominates for $p \to 1$.

c) The preferential attachment (12.7) gives a clear advantage to the "old" sides, and the fitness described above was designed to correct this "injustice". Another way of obtaining the same effect is to give an advantage to new sites by introducing in the dynamic equation (12.7) an "aging" factor τ_i for each site i, and assuming that the probability of connecting a new site with some old one is proportional not only to the connectivity of the old site but also to a power of its age, $\tau_i^{-\alpha}$. It turns out [165] that the power-law distribution is still obtained for $\alpha < 1$, while for $\alpha > 1$ the exponential law applies, with the intermediate case $P(k) = const$ for $\alpha = 1$. The "winner takes all" behavior occurs for large α, such as $\alpha > 10$.

12.5 Mean Field Analysis of Networks

In the previous section, we applied to the analysis of networks some ideas from the thermodynamic theory of equilibrium processes, such as a mean field type dynamic equation (12.7), and the "chemical potential" approach which led to Bose condensation. We now present an elegant general approach [166] based on the Landau mean field theory of phase transitions considered in Chapter 4. In order to do this, we need to characterize a network by the thermodynamic potential G which is a function of the order parameter η and the field H conjugate to it, $G = G(\eta, H)$, just as for the magnetic Ising system with order parameter M the free energy G is a function of M and the magnetic field H conjugate to M, $G = G(M, H)$. As usual, we require that the order parameter η vanishes in the disordered phase.

A site with connectivity k feels, on the average, k of its neighbors, i.e., it is subject an average field $k\eta$ in addition to the external field

H. Hence, the Landau expansion of $G(\eta, H)$ contains both of the fields H and $k\eta$, and can be written in the following form

$$G(\eta, H) = -\eta H + \sum_k P(k)\phi(\eta, k\eta),\qquad (12.22)$$

i.e., it depends on the distribution function $P(k)$. The following analysis [166] is based on the expansion of the function $\phi(x, y)$ in a Taylor series for small x and y,

$$\varphi(x, y) = \sum_m \sum_l \varphi_{ml} x^m y^l,\qquad (12.23)$$

and analyzing the first terms of this expansion. The coefficients in this expansion are of the form

$$\frac{d^n G(\eta, H)}{d\eta^n}\bigg|_{\eta=0} = n! \sum_{l=0}^n \varphi_{n-l,l} < k^l > .\qquad (12.24)$$

The average values $< k^l >$ entering Eq. (12.24) are completely determined by the distribution function $P(k)$. For the power law distribution, $P(k) \sim k^{-\gamma}$, we find that

$$< k^l > \equiv \int_1^\infty k^l P(k) dk \sim k^{l-\gamma+1}|_{k\to\infty},\qquad (12.25)$$

which means that all moments with $l \geq \gamma$ diverge. For these divergent moments, the function (12.22) contains singular coefficients (12.24), and, therefore, the Landau mean field approach breaks down. Just as for equilibrium phenomena, such a breakdown is caused by strong fluctuations. The most connected sites in the networks, which correspond to strong fluctuations in the local connectivity, are the source of the inapplicability of mean field theory. As the power γ in $P(k) \sim k^{-\gamma}$ is increased, the relative number of highly connected sites decreases, and the phase transition becomes of the mean field type. Calculations show [166] that this happens for $\gamma > 5$, while there is a logarithmic corrections for $\gamma = 5$. (This latter feature can be compared with the analogous behavior near the upper critical dimension $d = 4$ of the thermodynamic critical phenomena). Other results are that for $3 < \gamma < 5$, the critical indices are functions of γ

$(\eta \sim (|T - T_c|/T_c)^{\frac{1}{\gamma-3}}$ etc.), while no phase transition occurs at any finite temperature for $2 < \gamma \leq 3$. These results have been obtained in the absence of the third order term in η in the expansion of G. If this term is included, G contains singular terms of the form:

a) $\eta^3 \ln(\eta)$ for $\gamma = 4$,

b) $\eta^{\gamma-1}$ for $2 < \gamma < 3$ and $3 < \gamma < 4$,

c) $\eta^2 \ln(\eta)$ for $\gamma = 3$.

One further comment should be made. We saw in the previous chapter that for small world phenomena, a small fraction p of long-range bonds brings a system into the mean field universality class. However, for scale-free systems, in spite of the existence of the long-range bonds, the situation is quite different. As we have seen, the mean field behavior only takes place for large enough values of the power γ in the distribution function of k, $\gamma > 5$, while for smaller values of γ the critical behavior is model independent but non-universal, and the critical indices depend on γ.

12.6   Hubs in Scale-Free Networks

Scale-free networks with a power-law distribution function have a topological structure very different from that of usual ("exponential") networks where the distribution function has a Gaussian form. In the exponential networks nearly all sites have more or less the same number of links, close to the average one, and the system is homogeneous. The situation is quite different for scale-free networks. As we have seen, the dynamics of these networks ensures that "rich becomes richer" or even that the "winner takes all". This means that scale-free networks are very inhomogeneous, with a few sites ("hubs") having a very large number of links and most of the sites having only very few links, as shown in Fig. 12.1. By the way, such a structure is typical for the routes of large airlines which fly between many sites using the hubs as intermediate stations for connecting flights. The existence of hubs is very important for phase transitions, such as the ordering of spins located on sites of scale-free system. Since the hubs

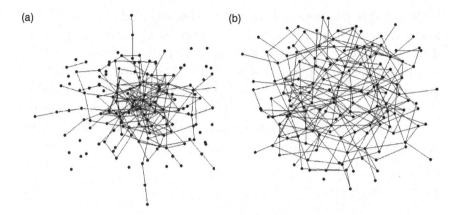

Fig. 12.1   Two networks containing 150 sites and 4950 links which are distributed among the sites according to (a) a power-law and (b) an exponential law. The former have a few sites with many links ("hubs") such as Chicago and Atlanta for air flights, or Google and Yahoo for the Internet.

have many links, they keep the majority of spins in an ordered state. To destroy the order or reverse the direction of the spins, it is enough to flip the spins located at a small numbers of hubs, some $10^{-3} - 10^{-4}$ of all the sites [167].

Another interesting consequence of the distinction between exponential and scale-free networks is their different response to damage [168]. Damage means the removal of some sites with all their links. A distinction must be made between random (failure) and intentional (attack) damage. While for the exponential networks there is no difference between these two types of damage since all sites are more or less equivalent, a striking difference exists for scale-free networks [169], [170]. It is intuitively clear that scale-free networks are extremely resilient to random damage but are very sensitive to intentional damage focused on the hubs, which destroys many links, and so interrupts the connection between different parts of the network.

A specific example of the above process, which is of great practical interest, concerns the stability of the Internet, which can be studied on the basis of percolation theory. The main question is how many sites need to be destroyed (or incapacitated) before the Internet becomes unstable, i.e., the connectivity of the world wide web is broken and it disintegrates into small disconnected clusters. It was found [169; 170] that the Internet network would never disintegrate if the system were infinite. While the Internet is a finite rather than an infinite network, it remains connected even if more than 90% of the sites are removed at random. However, if an attack is planned to eliminate preferentially the most connected sites (the hubs), even if less than 10% of these sites are removed the Internet will break up into disconnected sections.

12.7 Conclusion

Many natural and man-made systems belong to the class of scale-free networks, which are characterized by power-law distributions, $P(k) \sim k^{-\gamma}$, of the number k of connections between one object and others in these systems. The value of the exponent γ turns out to be sensitive to the details of the network structure, and it is usually in the range $2 < \gamma < 3$. There is no conclusive theory which either predicts the value of γ or classifies networks according to the magnitude of γ. In contrast to small world systems, where the mean field type transition occurs for all values of the characteristic parameter p, scale-free systems show a whole range of different types of phase transitions for different values of γ. This subject is developing very rapidly, and several reviews [171], [172], [165] have been published recently.

Bibliography

[1] L. Landau and E. Lifshitz, *"StatisticalPhysics"* (Pergamon, 1980).

[2] P. Weiss, J. Phys. **6**, 661 (1907).

[3] I. Prigogine and R. Defay, *"Chemical Thermodynamics"* (Longmans Green, New York, 1954).

[4] A. A. Galkin and Lunin V. V., *Russian Chem. Rev.* **74**, 21 (2005).

[5] S. N. V. K. Aki and Abraham, *Envir. Prog.* **17**, 246 (1998).

[6] S. N. V. K. Aki and Abraham, *Chem. Eng. Sci.* **54**, 3543 (1999).

[7] A. Kruse and H. Vogel, *Chem. Eng. Technol.* **31**, 23 (2008).

[8] M. N. da Ponte, *J. Supercrit. Fluids* **47**, 344 (2009).

[9] H. E. Stanley, *Rev. Mod. Phys.* **71**, S358 (1999).

[10] H. Boukari, J. N. Shaumeyer, M. E. Briggs, and R. W. Gammon, *Phys. Rev. A* **41**, 2260 (1990).

[11] H. Klein, G. Schmitz, and D. Woermann, *Phys. Rev. A* **43**, 4562 (1991).

[12] P. Guenoun, B. Khalil, D. Beysens, Y. Garnabos, F. Kammoun, and B. Le Neindre, *Phys. Rev. E* **47**, 1531 (1993).

[13] M. Bonetti, D. Perrot, and D. Beisens, *Phys. Rev. E* **49**, 4779 (1994).

[14] H. Boukari, R. L. Pego, and R. W. Gammon, *Phys. Rev. E* **52**, 1614 (1995).

[15] L. Onsager, *Phys. Rev.* **65**, 117 (1944).

[16] R. B. Potts, *Proc. Cambridge Philos. Soc.* **48**, 106 (1952).

[17] S. G. Brush, *Rev. Mod. Phys.* **39**, 883 (1967).

[18] J. M. Ziman, *Properties of the Theory of Solids* (Cambridge, 1964).

[19] C. Kittel, *Amer. J. Phys.* **37**, 917 (1969).

[20] X. Wang, "Physical Examples of Phase Transition in One-Dimensional Systems with Short Range Interaction", 2012.

[21] S. T. Chui and J. D. Weeks, *Phys. Rev. B* **23**, 2438 (1981).

[22] T. Dauxois, M. Peyrard, and A. R. Bishop, *Phys. Rev. Lett.* **47**, R44 (1993).

[23] S. Ares and A. Sanchez, *Eur. Phys. J. B* **56**, 253 (2007).

[24] N. M. Svrakic, *Phys. Lett. A* **80**, 43 (1980).

[25] B. Liu and M. Gitterman, *Am. J. Phys.* **71**, 806 (2003).

[26] K. Binder and E. Luijten, *Phys. Rep.* **344**, 179 (2001).

[27] P. G. de Gennes, *"Superconductivity of Metals and Alloys"* (Benjamin, 1966).

[28] B. Rosenstein and D. Li, *Rev. Mod. Phys.* **82**, 105 (2010).

[29] V. L. Ginzburg, *Sov. Phys. Solid State* **2**, 1824 (1960).

[30] K. Wilson, *Rev. Mod. Phys.* **47**, 773 (1975).

[31] K. Wilson, *Physica* **73**, 119 (1974).

[32] A. Hankey and H. E. Stanley, *Phys. Rev. B* **6**, 3515 (1972).

[33] G. S. Rushbrooke, *J. Chem. Phys.* **39**, 8942 (1963).

[34] R. B. Griffiths, *Phys. Rev. Lett.* **14**, 623 (1965).

[35] L. Kadanoff, *Physics* **2**, 263 (1966).

[36] H. J. Maris and L. P. Kadanoff, *Amer. J. Phys.* **46**, 652 (1978).

[37] G. Nicolis, *Introduction to Nonlinear Science* (Cambridge, 1995).

[38] Th. Niemeijer and J. M. J. van Leeuwen, *Physica* **71**, 17 (1974).

[39] E. M. Lifshitz and L. P. Pitaevski, *Statistical Physics*, Part 2 (Pergamon, 1980).

[40] B. G. Levi, *Physics Today*, December 1997, p. 17.

[41] W. Ketterle, *Scientific American*, December 1999, p. 30: K. Burnett, M. Edwards, and C. W. Clark, *ibid.* p. 37.

[42] E. A. Cornell and C. E. Wieman, *Rev. Mod. Phys.* **74**, 875 (2002).

[43] P. Meijer, *Amer. J. Phys.* **62**, 1105 (1994).

[44] L. N. Cooper, *Phys. Rev.* **104**, 1189 (1956).

[45] N. F. Mott, *Physica A*, **200**, 127 (1993).

[46] R. S. Knox, *"Theory of excitons"* (Academic Press, 1963).

[47] B. D. Josephson, *Phys. Lett.* **1**, 251 (1962).

[48] P. W. Anderson and J. M. Rowel, *Phys. Rev. Lett.* **10**, 230 (1963).

[49] J. G. Bednorz and K. A. Muller, *Z. Phys. B* **64**, 189 (1986).

[50] *Physica C* **385**, issue 1-2 (2003).

[51] P. J. Ford and G. A. Sanders, *Contemporary Physics* **38**, 63 (1997).

[52] M. Kac, G. E. Uhlenbeck, and P. C. Hemmer, *J. Math. Phys.* **4**, 216, 229 (1963).

[53] R. J. Baxter, *Ann. Phys. NY*, **70**, 193 (1972).

[54] F. W. Wu, *Phys. Rev. B* **4**, 2312 (1971).

[55] L. P. Kadanoff and F. J. Wegner, *Phys. Rev. B*, **4**, 3989 (1971).

[56] V. G. Vaks, A. I. Larkin, and Y. N. Ovchinnikov, *Sov. Phys. JETP* **22**, 820 (1965).

[57] M. Gitterman and P. Hemmer, *J. Phys. C* **13**, L329 (1980).

[58] R. J. Baxter and F. Y. Wu, *Austr. J. Phys.*, **27**, 357, 369 (1974).

[59] D. Imbro and P. C. Hemmer, *Phys. Lett. A*, **57**, 797 (1976).

[60] M. Gitterman and M. Mikulinsky, *J. Phys. C* **10**, 4073 (1977).

[61] E. E. Gorodetsky and M. A. Mikulinsky, *Zh. Eksp. Teor. Fiz.* **66**, 986 (1974) [*Sov. Phys - JETP* **39**, 480 (1974)].

[62] S. Alexander and D. J. Amit, *J. Phys. A* **8**, 1988 (1975).

[63] L. P. Kadanoff, *Phys. Rev. Lett.* **23**, 1439 (1969).

[64] R. J. Baxter, M. F. Sykes, and M. G. Watts, *J. Phys. A* **8** , 245 (1975).

[65] M. Kac, *Physics Today*, October 1964, p. 40.

[66] T. H. Berlin and M. Kac, *Phys. Rev.* **86**, 821 (1952).

[67] J. J. Binney, N. J. Dowrick, A. J. Fisher, and M. E. J. Newman, *The Theory of Critical Phenomena* (Clarendon, Oxford, 1992).

[68] M. D. Mermin and H. Wagner, *Phys. Rev. Lett.* **17**, 133 (1966).

[69] J. M. Kosterlitz, *J. Phys. C* **7**, 1046 (1974).

[70] G. Blatter, M. V. Feigelman, V. B. Geshkenbein, A. I. Larkin, and V. M. Vinokur, *Rev. Mod. Phys.* **66**, 1125 (1994).

[71] N. B. Kopnin, *Rep. Prog. Phys.* **65**, 1633 (2002).

[72] Ya. B. Zeldovich, *Zh. Fiz. Chim.* **11**, 685 1938 (in Russian).

[73] R. Aris, Arch. *Ration. Mech. Anal.* **19**, 81 (1965).

[74] R. Aris, Arch. *Ration. Mech. Anal.* **27**, 356 (1968).

[75] J. M. Powers and S. Pautucci, *Am. J. Phys.* **76**, 848 (2008).

[76] E. B. Starikov and B. Norden, *J. Phys. Chem B* **111**, 14431 (2007).

[77] A. L. Cornish-Bowden, *J. Bioscences* **27**, 121 (2002).

[78] M. Gitterman and V. Steinberg, *J. Chem. Phys.* **69**, 2763 (1978); *Phys. Rev. Lett.* **35**, 1588 (1975).

[79] M. Gitterman and V. Steinberg, *Phys. Rev. A* **20**, 1236 (1979).

[80] W. Ebeling and R. Sandig, *Ann. Phys. (Leipzig)* **28**, 289 (1973).

[81] Y. Albeck and M. Gitterman, *Phil. Mag. B* **56**, 881 (1997).

[82] H. S. Caram and L. E. Scriven, *Chem. Eng. Sci.* **31**, 163 (1976).

[83] G. Othmer, *Chem. Eng. Sci.* **31**, 993 (1976).

[84] L. R. Corrales and J. Wheeler, *J. Chem. Phys.* **91**, 7097 (1989).

[85] V. Talanquer, *J. Chem. Phys.* **96**, 5406 (1992).

[86] D. Borgisand and M. Moreau, *J. Stat. Phys.* **50**, 935 (1988).

[87] Y. Rabin and M. Gitterman, *Phys. Rev. A* **29**, 1496 (1984).

[88] H. Reiss and M. Shugard, *J. Chem. Phys.* **65**, 5280 (1976).

[89] M. Gitterman in *"Nonequilibrium Statistical Mechanics"*, ed. E. S. Hernandez [World Scientific, 1989], p. 103.

[90] Y. Albeck and M. Gitterman, *Phys. Rev. Lett.* **60**, 588 (1988).

[91] M. Gitterman, B. Ya. Shapiro, and I. Shapiro, *Phys, Rev. B* **65**, 174510 (2002).

[92] M. Gitterman, B. Ya. Shapiro, and I. Shapiro, *Europhys. Lett.* **76**, 1158 (2006).

[93] B. Kalisky, D. Giller, A. Shaulov, and Y. Yeshurun, *Phys. Rev. B* **67R**, 140508 (2003).

[94] J. L. Tveekrem and D. T. Jacobs, *Phys. Rev. A* **27**, 2773 (1983).

[95] R. H. Cohn and D. T. Jacobs, *J. Chem. Phys.* **80**, 856 (1984).

[96] D. T. Jacobs, *J. Chem. Phys.* **91**, 560 (1989).

[97] C. D. Specker, J. M. Ellis, and J. K. Baird, *Int. J. Thermophys.* **28**, 846 (2007).

[98] J. K. Baird and J. C. Clunie, *J. Phys. Chem.* **102**, 6498 (1998).

[99] J. Ke, B. Han, M. W. George, H. Yan, and M. Poliakoff, *J. Am. Chem. Soc.* **123**, 3661 (2001).

[100] B. Hu, R. D. Richey and J. K. Baird, *J. Chem. Eng. Data*, **54**, 1537 (2009).

[101] M. E. Fisher, *Phys. Rev.* **176**, 257 (1968).

[102] M. Gitterman and V. Steinberg, *Phys. Rev. A* **22**, 1287 (1980).

[103] M. A. Anisimov. A. V. Voronel, and E. E. Gorodetski, *Sov. Phys - JETP,* **33**, 605 (1971).

[104] J. Troen, R. Kindt, W. van Dael, M. Merabet, and T. K. Bose, *Physica A* **156**, 92 (1989).

[105] E. U. Frank, in *"The Physics and Chemistry of Aqueous Ionic Solutions"*, eds. M. C. Bellissent-Funel and G. W. Nelson [Reidel, Dordecht, 1987], p. 337.

[106] A. Stein and G. F. Allen, *J. Chem. Phys.* **59**, 6079 (1973).

[107] G. H. Shaw and G. F. Allen, *J. Chem. Phys.* **65**, 4906 (1076).

[108] E. M. Anderson and S. C. Greer, *Phys. Rev. A* **30**, 3129 (1984).

[109] D. Jasnow, W. I. Goldberg, and J. S. Semura, *Phys. Rev. A* **9**, 355 (1974).

[110] M. E. Fisher and J. S. Langer, *Phys. Rev. Lett.* **20**, 665 (1968).

[111] M. Gitterman, *Phys. Rev. A* **28**, 358 (1983).

[112] J. C. Wheeler, *Phys. Rev. A* **30**, 648 (1984).

[113] G. R. Anderson and J. C. Wheeler, *J. Chem. Phys.* **69**, 2082 3403 (1978); ibid. **73**, 5778 (1980).

[114] J. L. Tveekrem, R. H. Cohn, and S. C. Greer, *J. Chem. Phys.* **86**, 3602 (1987).

[115] J. L. Tveekrem, S. C. Greer, and D. T. Jacobs, *Macromolecules* **21**, 147 (1988).

[116] S. T. Milner and P. C. Martin, *Phys. Rev. A* **33**, 1996 (1986).

[117] J. M. Chomaz and A. Couairon, *Phys. Fluids,* **11**, 2977 (1999).

[118] Y. Kuramoto and T. Tsuzuki, *Progr. Theor Phys.* **52**, 1399 (1974).

[119] A. Saul and K. Showalter in *"Oscillations and Travel Waves in Chemical Systems"*, ed. by R. J. Fieldand and M. Burger (Wiley, New York, 1985).

[120] R. Graham and H. Haken, *Z. Phys.* **237**, 31 (1970).

[121] F. Heslot and A. Libchaber, *Phys. Scr. T* **9**, 126 (1985).

[122] J. Lin and P. Kahn, *J. Math. Biol.* **13**, 383 (1982).

[123] M. Gitterman, *Phys. Rev. E* **70**, 036116 (2004).

[124] Y. Paltiel, E. Zeldov, Y. N. Myasoedov, H. Shtrikman, S. Bhattucharya, M. J. Higgins, E. A. Andrei, P. L. Gamme, and D. J. Bishop, *Nature (London)* **403**, 398 (2000).

[125] M. Golosovsky, M. Tsindlekht, H. Chaet, and D. Davidov, *Phys. Rev. B* **50**, 470 (1994).

[126] R. E. Hunt and D. G. Crighton, *Proc. Royal Soc (London), A* **435**, 109 (1991).

[127] G. L. Dolan, G. V. Chandrashekhar, T. R. Dingel, C. Feild, and F. Holzberg, *Phys. Rev. Lett.* **62**, 827 (1989).

[128] E. T. Whittaker and G. N. Watson, *"A Course of Modern Analysis"* (Cambridge University Press, 1927).

[129] R. C. Bourret, H. Frish, and A. Pouquet, *Physica* **65**, 303 (1973).

[130] P. Colet, D. Walgraef, and M. San Miguel, *Eur. J. Phys. B* **11**, 517 (1992); W. van Saarlos, *Phys. Rev. A* **39**, 6367 (1999).

[131] J. M. Chomaz, *Phys. Rev. Lett.* **69**, 1931 (1992).

[132] V. A. Vyssotsky, S. B. Gordon, H. L. Frisch, and J. M. Hammersley, *Phys. Rev.* **123**, 1566 (1961).

[133] J. M. Ziman, *J. Phys. C* **1**, 1532 (1968).

[134] J. W. Esssam, *Rep. Prog. Phys.* **43**, 833 (1980).

[135] R. B. Griffiths, *Phys. Rev. Lett.* **23**, 17 (1969).

[136] C. Kittel, *Introduction to Solid State Physics* (Wiley, New York, 1996).

[137] M. Mezard, G. Parisi, and M. Virasoro, *Spin Glasses Theory and Beyond* (World Scientific, Singapore, 1987).

[138] V. Dotsenko, *Introduction to the Replica Theory of Disordered Statistical Systems* (World Scientific Lecture Notes in Physics, Vol. 54, 2000).

[139] P. Larranaga, C. M. H. Kuijpers, R. H. Murga, I. Inza, and S. Dizdarevic, *Artificial Intelligence Review* **13**, 129 (1999).

[140] J. J. Hopfield, *Proc. Nat. Acad. Sci.* (USA) **79**, 2554 (1982) and **81**, 3088 (1984).

[141] S. Milgram, *Psychol. Today* **2**, 60 (1967).

[142] D. J. Watts and S. H. Strogatz, *Nature* **393**, 440 (1998).

[143] D. J. Watts, *Small World* (Princeton, New Jersey, 1999).

[144] S. N. Dorogovtsev and J. F. F. Mendes, *Adv. Phys.* **51**, 1079 (2002).

[145] M. E. J. Newman, *J. Stat. Phys.* **101**, 819 (2000).

[146] V. Latora and M. Marchiori, *Physica A* **314**, 109 (2002); *Eur. Phys. J. B* **32**, 249 (2003).

[147] M. E. J. Newman and D. J. Watts, *Phys. Lett. A* **263**, 341 (1999).

[148] A. Barrat and M. Weigt, *Eur. Phys. J. B* **13**, 547 (2000).

[149] M. Gitterman, *J. Phys. A* **33**, 8373 (2000).

[150] H. Hong, B. J. Kim, and M. Y. Choi, *Phys. Rev. E* **66**, 018101 (2002).

[151] B. J. Kim, H. Hong, G. S. Jeon, P. Minnhagen, and M. Y. Choi, *Phys. Rev. E* **64**, 056135 (2001).

[152] C. P. Herrero, *Phys. Rev. E* **65**, 066110 (2002).

[153] P. Sen, K. Banerjee, and T. Biswas, *Phys. Rev. E* **66**, 037102 (2002).

[154] V. Pareto, *Le Cours dE'conomie Politique* (MacMillan, London, 1897).

[155] Z. Burda, D. Johnston, J. Jurkiewicz, M. Kaminski, M. A. Nowak, G. Papp, and I. Zahed, *Phys. Rev. E* **65**, 026102 (2002).

[156] B. Gutenberg and C. F. Richter, *Seismity of the Earth and Associated Phenomena*, 2nd ed. (Princeton, NJ, Princeton University Press, 1954)

[157] G. K. Zipf, *The Psycho-biology of Language* (Houghton Mifflin Co., Boston, 1935).

[158] A. Y. Abul-Magd, *Phys. Rev. E* **66**, 057104 (2002).

[159] M. Gitterman, *Rev. Mod. Phys.* **50**, 85 (1978).

[160] G. Caldarelli, A. Capocci, P. De Los Rios, and M. A. Munoz, *Phys. Rev. Lett.* **89**, 258702 (2002).

[161] A.-L. Barabasi and R Albert, *Science* **286**, 509 (1999).

[162] G. Bianconi and A.-L. Barabasi, *Phys. Rev. Lett.* **86**, 5632 (2001).

[163] P. L. Krapivsky, S. Redner, and F. Leyvraz, *Phys. Rev. Lett.* **85**, 4629 (2000).

[164] Z. Liu, Y.-C. Lai, N. Ye, and P. Dasgupta, *Phys. Lett. A* **303**, 337 (2000).

[165] S. N. Dorogovstev and J. F. F. Mendes, *Phys. Rev. E* **62**, 1842 (2000).

[166] A. V. Goltsev, S. N. Dorogovtsev, and J. F. F. Mendes, *Phys. Rev. E* **67**, 026123 (2003).

[167] A. Aleksieyuk, J. A. Holyst, and D. Stauffer, *Physica A* **310**, 260 (2002).

[168] R. Albert, H. Jeong, and A.-L. Barabasi, *Nature*, **406**, 378 (2000).

[169] R. Cohen, K. Erez, D. ben-Avraham, and S. Havlin, *Phys. Rev. Lett.* **85**, 4626 (2000).

[170] R. Cohen, K. Erez, D. ben-Avraham, and S. Havlin, *Phys. Rev. Lett.* **86**, 3682 (2000).

[171] A.-L. Barabasi, and E. Bonabeau, *Sci. Am.* **288**, 60 (2003).

[172] R. Albert and A.-L. Barabasi, *Rev. Mod. Phys.* **74**, 47 (2002).

Index